W9-BBL-498

CATS

CATS

SMITHSONIAN ANSWER BOOK

JOHN SEIDENSTICKER AND
SUSAN LUMPKIN

PHOTOGRAPHS BY
ART WOLFE

SMITHSONIAN BOOKS
WASHINGTON

© 2004 by John Seidensticker and Susan Lumpkin
Photographs © 2004 by Art Wolfe

Copy editor: Robin Whitaker
Production editor: Joanne Reams
Designer: Janice Wheeler

Library of Congress Cataloging-in-Publication data:
Seidensticker, John.
Cats : Smithsonian answer book / John Seidensticker and Susan Lumpkin ;
    photographs by Art Wolfe.
        p.   cm.
    Includes bibliographical references and index.
ISBN 1-58834-125-9 (alk. paper)—ISBN 1-58834-126-7 (pbk. : alk. paper)
    1. Felidae—Miscellanea. I. Lumpkin, Susan. II. Smithsonian Books (Publisher). III. Title.
QL737.C23S4252   2004
599.75—dc22   2004041671

British Library Cataloging-in-Publication data available

Printed in Singapore, not at government expense
10 09 08 07 06 05 04   1 2 3 4 5

♾ The paper used in this publication meets the minimum requirements of the American National
Standard for Information Sciences—Permanence of Paper for Printed
Library Materials ANSI Z39.48-1992.

Frontispiece: **Requiring huge expanses of African plains habitat to survive, lions face an
uncertain future on that crowded continent. Human activities are squeezing out most other cat
species too. Greater awareness of the danger cats face is the first step toward our securing their
future in the wild.**

CONTENTS

.1.
CAT FACTS

.2.
CAT EVOLUTION AND DIVERSITY

.3.
CATS AND HUMANS

PREFACE

From the vantage point of our garden patio each summer night, we watch a half dozen or so domestic cats, some feral, some with collars, slowly patrol our alley just off Fourteenth Street in northwest Washington, DC. Some walk by a few minutes apart. Now and then, two will stroll by together. Many usually pause briefly at the gate and check if we are sitting out or not. Our garden pond with its goldfish and our garden birds are a big attraction when we are not around. We have been fascinated with cats, watching them for our professional careers and even before that, but we are reminded each evening of Niko Tinbergen's lament in *The Herring Gull's World:* "Even an hour's careful observation of the goings-on in a gullery faces one with a great number of problems—more problems, as a matter of fact, than one could hope to solve in a lifetime" (Tinbergen 1970, xiv). So it is with cats.

People love cats, or respect them, or loathe them, or are terrified of them, depending on cat and circumstance. Cats have been used as a model biological system to help us better understand disease, morphology, physiology, and behavior in various other species. We have studied cats' neuro-function to better understand how animals perceive the world. Biologists have teased apart the social lives of cats to help us better understand complex cooperative and competitive strategies. How the presence or absence of different wild cat species alters the composition of local ecological communities and the patterns of biological diversity across broad geographical ranges is the focus of a growing number of biological reports. We are learning how human-caused stresses on ecological communities result in the decline of populations of most species of wild cats. We sense that, as a result of this scientific activity, the awe, fear, and mystery that once surrounded wild cats are rapidly being replaced with appreciation. Our e-mail indicates that people have increasing numbers of questions about cats in general and about living with cats. We wrote this book to enable people to look at cats more closely, ask questions, and find accurate answers to their questions so that they can better enjoy and appreciate cats.

This book is arranged like the others in the Smithsonian Answer Book series. We start with "Cat Facts," a chapter that answers questions about cat biology, physical features, senses, diet and predation, social behavior, and cat life. Chapter 2, "Cat Evolution and Diversity," discusses the origin of cats, cat evolution, the diversity in cats, including descriptions of all the cat lineages and of all 40 species, and an explanation of how more than one wild cat species can live in a particular area. Chapter 3, "Cats and Humans," answers questions about the decline and recovery of wild cat populations, the domestication of cats and their various relationships with humans, cats and human culture, cat attacks on humans, and cats and science. Appendix 1 provides common and scientific names of the cat species, appendix 2 outlines information on the conservation status of wild cat populations, and appendix 3 lists some Web sites that offer more information about cats. The references include both general works on cats and specialized scientific references on individual topics. Art Wolfe's spectacular photographs bring even the rarest cats up close. Please note that wherever "John" is mentioned in the book, this refers to John Seidensticker, one of the two coauthors.

ACKNOWLEDGMENTS

We prefer our cats wild. Our positions at the Smithsonian's National Zoological Park and Friends of the National Zoo have allowed an unbroken and unparalleled opportunity to watch wild cats every day and to travel frequently to see and study wild cats where they live in the wild. This has been a splendid place from which we can share our passion for cats through our writing, exhibits, and lectures. Peter Cannell, late director of Smithsonian Institution Press, asked us to write this Smithsonian Answer Book. Vincent Burke encouraged and supported us while we were deeply involved in other projects. Nicole Sloan and Caroline Newman of Smithsonian Books saw the project to completion. Robin Whitaker's copy editing was masterful and much appreciated. Alvin Hutchinson tracked down dozens of papers and books in short order. Thank you.

We are especially grateful for the support John has received over the years from Friends of the National Zoo, Smithsonian's National Zoological Park, Smithsonian Special Foreign Currency Fund, World Wildlife Fund–US, World Wide Fund for Nature, ExxonMobil Foundation, National Fish and Wildlife Foundation, and Save The Tiger Fund and the generous hospitality of Estación Biológica de Doñana. This allowed John to visit and study cats where they live in the eastern deciduous forest and intermontane West of the United States, the Russian Far East, the mountains of central China, Thailand, Bangladesh, Nepal, India, Sri Lanka, Kenya, Spain, Yucatán in Mexico, and Sumatra, Java, and Bali in Indonesia. We thank our friends and colleagues who have shared their thoughts about cats with us over the years; they include Ted Bailey, Warren Brockleman, Kathy Carlstead, Ravi Chellam, Raghu Chundawat, John Eisenberg, Louise Emmons, Lyn DeAlwise, Eric Dinerstein, Nestor Fernandez, Joe Fox, Al Gardner, Josh Ginsburg, John Gittleman,

Helmut Hemmer, Rafael Hoogesteijn, Maurice Hornocker, Peter Jackson, Rod Jackson, A. J. T. Johnsingh, Warren Johnson, Ullas Karanth, Kae Kawanishi, Devra Kleiman, Margaret Kinnaird, Gary Koehler, Boonsong Lekagul, Paul Leyhausen, Chuck McDougal, Jeff McNeely, Jill Mellen, Hemanta Mishra, Sriyanie Miththapala, Fumi Mizutani, Dale Miquelle, Tim O'Brien, Steve O'Brien, Gustav Peters, Pat Quillen, Howard Quigley, Alan Rabinowitz, Douglas Richardson, George Schaller, Dave Smith, Alan Shoemaker, Mel and Fiona Sunquist, Ir. Suyono, Blaire Van Valkenburgh, Wang Sung, Wang Zong-Yi, Stuart Wells, Chris Wemmer, Wilber Wiles, and Eric Wikramanayake. Many others, wherever we have been and worked, have generously shared observations. This book is much better for their contributions. Any errors, of course, are ours.

The Smithsonian's mission, "the increase and diffusion of knowledge," is as exhilarating as it is challenging. A wise editor taught us that the first step is writing so that someone who cares will read and understand what you write. We wrote this book for our daughter, Lesley Seidensticker, and for our parents and their memory—Tom and Mary Lou Lumpkin and John and Gladys Seidensticker. This book is dedicated to them.

INTRODUCTION

Even young children can identify a cat as a cat, whether it is a tiny black-footed cat (*Felis nigripes*) or a tremendous tiger, so similar in form are these carnivores. However, the diversity of species in the family Felidae and the relationships among them are a source of continuing wonder and study for specialists. Officially, the number of cat species is now 40, up from 36 a decade ago and 37 two decades ago. This number has fluctuated because of taxonomic reclassifications. Within limits, morphological analysis has helped biologists advance our understanding of the evolutionary relationships of cats. However, biochemical methods are now providing further resolution, which has determined eight lineages, or clades, among the 40 cat species and has enabled relationships among the modern cats' ancestors to be traced back more than 10 million years. Modern field studies of most of the large and medium-sized species have been undertaken; however, facts about the small species remain virtually unknown. In this book, we have tried to bring together all that we do know.

Cat species constitute about 16 percent of the carnivores and fewer than 1 percent of all mammals. But the public's interest in and admiration for cats well exceed their proportional representation among the mammals. There are hundreds of millions of domestic cats but fewer than 100 Iriomote cats (*Prionailurus iriomotensis*) and 50 Iberian lynx (*Lynx pardinus*). Lions, tigers, pumas, and leopards, the very large and medium-sized cats, can be seen in most zoos, although most of the smaller wild cats are not found there. Some species are represented by only a few specimens in the world's museums.

Cats, which are highly specialized meat-eaters, range in size from 1 to 300 kilograms and appear at first glance to be variations on a common theme. On closer look, they are exquisitely intricate in their adaptation to place and circumstance. Most species are solitary, such as tigers and servals, although lions and cheetahs live in groups; all, however, have complex social lives. Cats live naturally on all continents but Antarctica and Australia; no cats have evolved there or on certain islands. However, by now people have taken domestic cats nearly everywhere. No cat species live on the ice or tundra or in the water, but they live nearly everywhere else. A strain of domestic cat has even adapted to living in freezers on freezer-living rats. One or more wild cat species live in habitats from true desert to mangroves and in nearly every terrestrial habitat in between.

Some cat species hunt in trees; some on the ground; and some on riverbanks, totally immersing themselves to catch fish. Wild cats hunt other animals as prey, the size of the cat species determining whether those prey are tiny insects or wild cattle weighing 1,000 kilograms. What they all have in common is the need for adequate prey and cover, which is also determined by the size of the species as well as its special adaptations. Cats' position at the top of the food chain is easily disrupted. Populations of most wild cat species are now threatened with extinction. Seeing their beauty and discovering their diversity and life processes are essential steps toward saving these cats.

.1.

CAT FACTS

WHAT ARE CATS?

Cats are vertebrates belonging to the class Mammalia. Like humans and about 4,630 other mammal species, from bats to whales, cats are warm-blooded, have hair, and feed their young with milk produced by mammary glands (this is the origin of the word *mammal*). These are the defining traits of mammals, but almost all mammals share other features too. Cats and all but a few other mammals give birth to live young. The senses of all mammals—hearing, vision, taste and smell, balance, and touch—are similar, although sensory abilities vary among groups and species. For instance, cats, unlike humans, get around in the dark by relying on their long sensitive whiskers to help them sense the objects in their environment. On the other hand, cats don't experience vivid colors as humans do; they see the world in shades of gray or, at most, in muted blues and greens.

Cats belong to the order of mammals called the Carnivora, a word that means "meat-eater" (see the following table and Figure 1). This order includes the family Felidae (cats) as well as other families that share characteristics designed for a diet of meat (about 15 families, although the number keeps changing). The Carnivora order is divided into two groups: the catlike carnivores (suborder Feliformia) and the bearlike carnivores (suborder Caniformia). The bearlike carnivore families include dogs and wolves (Canidae), bears (Ursidae), walrus (Odobenidae), fur seals

Scientists coined the term *hypercarnivore* to describe jaguars (the cat pictured on the facing page) and other cats, recognizing their near-complete reliance on diets of meat from animals they hunt and kill. This characteristic determines what cats look like (both inside and out), where they live, and how they relate to others of their species.

1

Empire Eukaryota
    Kingdom Metazoa (animals)
        Phylum Chordata (chordates)
            Subphylum Vertebrata (vertebrates)
                Class Chondrichthyes (cartilaginous fishes)
                Class Osteichthyes (bony fishes)
                Class Amphibia (frogs and salamanders)
                Class Reptilia (snakes, lizards, and relatives)
                Class Aves (birds)
                Class Mammalia (mammals) (29 orders, including):
                    Order Carnivora[†] (15 families, including):
                        Family Felidae[‡]

[†] See Figure 1 (opposite) for the relationships among carnivores.
[‡] See Figure 4 (page 134) for the relationships among the cats.

and sea lions (Otariidae), earless seals (Phocidae), raccoons (Procyonidae), weasels and otters (Mustelidae), and skunks (Mephitidae). The catlike families include civets (Viverridae), mongooses (Herpestidae), hyenas (Hyaenidae), and cats. Among these, cats' nearest genetic relatives are hyenas.

A key anatomical distinction between the catlike and bearlike carnivores lies in the auditory bulla, which is the part of the skull between the eardrum and the base of the outer ear. Occurring on each side of the skull and housing the ossicle of the middle ear, this bony chamber is filled with air and serves as a resonating chamber. In some species of both groups of carnivores, the auditory bullae may be inflated, which increases hearing sensitivity. The catlike carnivorans possess a unique arrangement in the auditory bullae. Each bulla is divided by a septum that varies in depth in different species and enables them to hear with greater acuity the sound frequencies that match the calls of their primary prey. This septum is absent from the bullae of the bearlike carnivorans.

Some species in the carnivore order supplement their meat diet with plant material, and some, such as giant pandas, eat mostly plant material. Cats, however, with the rarest exception, eat only meat. For this reason, scientists have coined the term *hypercarnivore* to describe cats. So, despite their range in mass from about 1 kilogram to 300 kilograms, all cats are adapted to hunt for live, mostly vertebrate prey (animals with backbones). As a result, cats form a remarkably homogeneous group.

HOW ARE CATS CLASSIFIED?

A species consists of a number of individual animals with very similar features. Males and females within a species are able to breed and produce viable young that

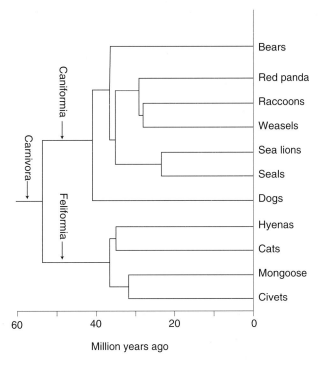

Figure 1. Combining morphological and genetic characteristics, scientists have created a tree that reflects evolutionary relationships among the Carnivora. The tree clearly shows the distinct split between the catlike carnivorans and the bearlike carnivorans. (Adapted from Bininda-Emonds, Gittleman, and Purvis 1999)

are fertile; under typical natural conditions, they do not breed with members of other species, although some exceptions occur.

The 2004 edition of *Mammals of the World*, an authoritative source, now lists 40 species in the family Felidae; the previous edition listed 36. As more is learned, these numbers will continue to change. For instance, the 2004 edition separates one species, the South American pampas cats, into three species (*Leopardus braccatus*, *L. colocolo*, and *L. pajeros*), but some scientists believe that another South American species, the oncilla (*Leopardus tigrinus*), should similarly be divided into two on the basis of genetic evidence. Others suggest that both the tiger and the clouded lepoard could be split into two species.

Many species arise as a result of geographic isolation, so some isolated populations may be on the borderline between subspecies and species. Scientists do not always agree on whether animals in two different populations are sufficiently different to be classified as separate species. In its classification the Iriomote cat, which lives on a single island near Taiwan, has flipped back and forth between the status of a full species and that of a subspecies of the widespread leopard cat. In some cases, we lack the information needed to resolve the issue. For this reason, different experts may list different numbers of species. For the cats, many of the species are well-defined, but relationships among them are not, although these are becoming known, with most cat species now assigned to one or another of several "lineages" of related species (see *When and Where Did Cats Evolve?* and *What Characterizes the Lineages of Cats?*).

HOW DO CATS DIFFER FROM OTHER CARNIVORES?

The species of the order Carnivora—about 250 of them—are not the only meat-eating mammals. Virginia opossums, frog-catching bats, squid-eating sperm whales, and many other, very different mammalian species share a taste for flesh. What then unites cats and dogs but separates cats and bats? The flat, shearlike blades on the fourth premolars of the upper jaw and the first molars of the lower jaw define the carnivores. Although in some members of this order, such as bears, these teeth, called carnassials, have lost the shear blades, in cats they have remained the ideal shape for slicing meat. Similarly, a relatively short digestive system, resulting from the easy digestion of meat, is characteristic of carnivores. Hypercarnivores, which eat only meat, have shorter digestive systems than carnivores that include other items in their diets. A puma, for instance, has a gut length about four times its body length, while the red fox, whose diet includes fruits, nuts, and insects, has a gut length five times its body length, plus a cecum (an extension of the large intestine).

Among the carnivores the cats are the most dramatically adapted for a predatory lifestyle. They are sleek runners and stealthy ambush hunters, possessing a unique combination of lethal, stabbing teeth, powerful jaws, flesh-ripping razor claws, strong agile bodies with flexible limbs, and excellent binocular vision.

HOW ARE CATS ALIKE?

There are certain things in nature in which beauty and utility, artistic and technical perfection, combine in some incomprehensible way: the web of a spider, the wing of a dragonfly, the superbly streamlined body of a porpoise, and the movements of a cat.

—Nobel laureate Konrad Lorenz, *Man Meets Dog*

For all their diversity in coat patterns, size, and habit, the felids are remarkably uniform in body and skull shape and proportions. Mammalogist John Eisenberg (1986) described the study of the Felidae as the study of variation on a common theme. A cat has a perfect body for its life as a predator. Skeletal and other anatomical differences among cats are almost entirely in size and, to a lesser extent, proportion (see Figure 2 and *What Are the Biggest Cat and the Smallest Cat?*).

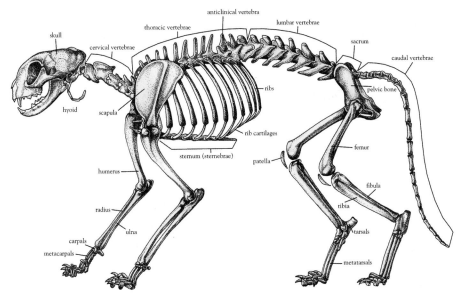

Figure 2. The skeletons of cats exhibit variation on a common theme, with only the size and relative proportions of the various bones varying much from species to species. Characteristics of the skull, such as length and breadth of the canine teeth, are related to the diet of the particular species, while postcranial traits, such as limb length, vary with habitat-utilization patterns of the species. (From Seidensticker and Lumpkin 1991, © Weldon Owen Pty Ltd)

The Genes

Stephen O'Brien of the National Cancer Institute and his associates have studied the genetics of domestic cats for more than 20 years. They have also explored the genetics of many species of wild felids. Their studies reveal that the genomes of most of the wild cat species and those of the domestic cats are nearly identical; of the 19 chromosome pairs that most cats possess, 15 are invariant among all of the species (summarized in O'Brien et al. 2002). Cats in the mostly South American Leopardus Lineage, such as ocelots and margays, differ from all others in having 18 pairs of chromosomes. Cats have roughly 50,000 genes, about the same number as humans do.

The Head

A cat's skull has short, powerful jaws, with attachments for large jaw muscles, which are needed to kill and eat the animals it preys on. Natural selection has optimized the design of a cat's jaw for a reasonably wide gape and a powerful killing bite and cutting action. The shorter jaw length gives extra strength to the temporalis, the muscle that closes the jaw. This muscle is attached to the side of the braincase. Interestingly, the skulls of many small cats have room for this large muscle to

This clouded leopard skull reveals the typical features of felid skulls. It is short, robust, and has large eye sockets placed well forward on the face. As in many cats, a ridge of bone, called a sagittal crest, runs along the top of the skull. Relative to its body size, however, the clouded leopard's canine teeth are the largest of all cats.

attach at the sides, but not the skulls of bigger cats. Large cat species and males in medium-sized species have bony ridges, called sagittal crests, along the top of the skull, providing the necessary surface area for this muscle to attach and creating even greater jaw strength. The tiny flat-headed cat, although one of the smaller cats, also has a well-developed crest, suggesting that this fish-eater has great biting power.

Teeth

Most cats have 30 teeth, but some short-faced cats (the four species of lynx, the Asian golden cat, African golden cat, and the caracal) have 28. The dental formula, on each side of the jaw, is three upper and three lower incisors, one upper and one lower canine, and two or three upper premolars and two lower premolars, and two upper and two lower molars.

The teeth are specialized for different purposes. Long, stabbing canines are used for the bite that kills prey. In general the canine teeth of cats are broader and more robust than those of other carnivores. Bladelike premolars (the carnassials) on the upper jaw and molars on the lower jaw are designed like scissors for shearing meat

Long, strong canine teeth, coupled with short, powerful jaws, enable pumas and other cats to kill prey with a single bite. Cheek teeth are used to slice meat, while the relatively small front teeth perform such delicate functions as plucking feathers from a bird carcass.

off bones. Incisors help grab hold of prey and do fine work such as plucking feathers or fur. Cats cannot move their jaws laterally and hence do not chew; they cut and tear their food into chunks they can swallow.

Like other mammals, cats have deciduous teeth. This means that cats first have a set of "milk teeth," what we usually call our baby teeth. These teeth erupt a week or two after birth in domestic cats. Milk teeth are gradually replaced by permanent teeth but in such a way that cats are never without canines or carnassials. However, for a short period of time, while the roots of the milk canines are disappearing and those of the permanent ones are still developing, the canines do not work. During this interval, young cats must rely on their mothers for food; even though they are capable of hunting alone, they cannot kill their prey. The age at which the permanent teeth are fully erupted varies among cat species, roughly by their size. A small

Cats have very flexible spines, an adaptation that increases running speed. While running, a cat's belly muscles tighten, making the spine arch like a taut bow, as in this lion (left). The relaxation of the muscles, like the release of a bow string, creates the energy for the explosive next step. The flexion of the spine that accompanies this is most pronounced in the speedy cheetah (right). Extended claws help a cat control the movements of captured prey so that the cat can deliver a killing bite. Cats also claw other objects to keep these vital weapons sharp, but we appreciate this behavior far more when it is performed by a lion in the wild than by a cat in our living room.

wildcat's (*Felis silvestris*) permanent teeth are in place by about 3.5 months; those of a tiger, by 12 to 18 months.

Posterior Anatomy

Most of a cat's skeleton is designed for speed and power. Cats have long legs and even walk on their toes to extend the length of their limbs (the toe shoes worn by ballet dancers act in the same way). Long legs and the "digitigrade" stance increase stride length, or the amount of ground covered with each step. Another adaptation that increases cats' stride length is their reduced or missing collar bone (clavicle), which allows freer movement of the front legs. Also the shoulder blades are free from the shoulder joints, so they further extend forelimb length.

The spine is very flexible because the articulations of the vertebrae are smooth and rounded, allowing rotation along the length of the spine. As a result, cats can flex and arch the back to increase speed. This also enables a cat to twist and turn easily. Overall, a cat's body is slender and fairly elastic, allowing it to move smoothly and gracefully. A long flexible tail, with from 14 to 28 caudal vertebrae in most species, acts like a rudder to improve balance.

The hind limbs of cats are generally longer than the forelimbs, an important feature for acceleration and jumping, because hind limbs provide more force than the forelimbs. Particularly long hind limbs are therefore expected in fast runners and

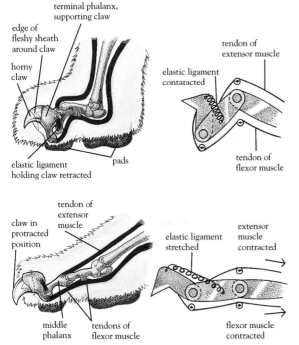

Figure 3. When a felid's claws are retracted, the spring ligament is contracted and the dorsal and ventral muscles are relaxed. This is the normal state of the claws when a cat is at rest. In use, the dorsal and ventral muscles contract, and the spring ligament stretches to extend the claws beyond their sheaths. Many carnivores, and all cats, possess retractile claws that function like this. (From Seidensticker and Lumpkin 1991, © Weldon Owen Pty Ltd)

good leapers. Curiously, cheetahs, which accelerate very rapidly and are the fastest runners among the felids, have hind limbs about as long as, or just slightly longer than, those of other cats of their size. Their thigh muscles, however, are 50 percent heavier than those of a typical mammal of the same weight. They also have an exceptionally flexible backbone, which contributes to their long stride length. Hind limb length is also related to prey-capture method; for instance, servals, which have very long hind limbs, pounce on their prey.

Paws and Claws

All cats have five claws on their forefeet and four on their hind feet. The claw on the first digit of the forefoot (the digit equivalent to our thumb), called the dewclaw, is very small and does not touch the ground. Cheetahs are exceptional in this regard: the dew claw on their forefoot is long and is used to snag and drag down prey. Cats use their claws to help grab and cling to prey, to climb trees, and as weapons in fights with other cats. Their claws can close around objects, as our fingers can. Although claws are useful for climbing up trees, acting like a mountain climber's pitons, they are less useful for climbing down (see *How Do Cats Climb?*).

To keep these tools sharp, a ligament holds the claws inside a protective sheath of skin except when they are in use. When needed, the claws are actively extended, or protracted, with a muscle contraction that stretches the springlike ligament. The mechanism is similar to the one that opens a jackknife (see Figure 3).

The deadly sharp claws of lions and most cats are usually hidden, retracted into sheaths on the toes to protect them from wear and tear. When needed to snare prey, climb a tree, or lash out at a competitor, the claws spring out like opening jackknives.

Extended claws help a cat control the movements of captured prey so that the cat can deliver a killing bite. Cats also claw other objects to keep these vital weapons sharp, but we appreciate this behavior far more when it is preformed by a lion in the wild than by a cat in our living room.

The bottom of a cat's paw has soft cushionlike pads, one in the center and one at the tip of each toe. These cushion the feet when the cat is running, and their ridges may provide traction for stopping. The pads of cheetahs are particularly hard, perhaps because they run on hard ground.

Cats living in both very hot and very cold climates, such as sand cats living in deserts and lynx living in boreal forests (*Lynx lynx* and *L. canadensis*), tend to have fur-covered foot pads to insulate the feet from extreme surface temperatures. Wide, furry paws also spread the cat's weight, like snowshoes, making it easier to travel on both snow and sand. Snow leopards have huge paws to grip rocks in their mountain homes and can walk easily through deep snow, an ability shared by few other cats besides lynx. More arboreal cats, such as clouded leopards, also have very wide

This resting cheetah's claws are retracted but still visible. The sheathes that conceal the claws in most other cats are not present in cheetahs. In a few other species, such as fishing cats, the sheathes are relatively short, leaving part of the claws always exposed.

paws, perhaps better to grip large branches. The paw pads of marbled cats, which live in Southeast Asian forests and are believed to be arboreal, are twice the size of the pads of leopard cats (*Prionailurus bengalensis*), even though the two are otherwise about the same size.

WHY CAN'T CHEETAHS RETRACT THEIR CLAWS?

In fact, cheetahs can slightly retract their claws, but they are always exposed and visible because they lack the skin sheathes that usually conceal the retracted claws of other cats. This visibility has led many people to believe that cheetahs cannot retract their claws. For cheetahs, claws give traction while in hot pursuit of prey, just as the spikes on the shoes of runners give them traction. The claws of flat-headed cats and fishing cats also are always visible because of reduced sheathes, which leave about two-thirds of the claw bare. Neither do the sheathes of Geoffroy's cats cover their claws entirely.

HOW DO CATS CLIMB?

All cats can climb trees; one folktale claims that a cat's only trick is to climb a tree. Tigers, for instance, have been observed to climb trees to escape floodwaters. However, most adult tigers and lions are too heavy to climb easily and do so only when absolutely necessary. Climbing is also difficult for cheetahs, because their tibia and fibula are bound together by fibrous tissue, an adaptation that gives stability to the leg. This stability improves their locomotion, but in turn it reduces their ability to rotate their legs or feet to grasp a branch, as is the case for dogs.

The anatomical features that let cats climb are similar to those used to bring down and hold struggling prey. As described by Alan Turner, "A leopard climbing . . . proceeds with a series of bounds in which both forelimbs and both hind limbs act together. Lateral movements of the forelimbs are important, permitting the trunk to be grasped, while the hind limbs continue to move in line with the body. The muscles controlling flexion and extension of the spine are also very important during climbing" (Turner 1997, 146). To climb up a tree, cats rely on strong back muscles, strong hind feet, and splayed front toes tipped with curved claws that act as pitons to pull them up along the tree trunk or limbs.

The hard part is climbing back down, as anyone who's seen a domestic cat stuck in a tree knows. Leopards can run down trees, but, with the exception of margays and clouded leopards, most cats cannot climb down headfirst, as tree squirrels do. For one thing, the ankles and toes on the hind limbs are less mobile, so most cats cannot grasp a limb with the hind paws. Neither can the claws help in descending head first, because they are curved in the wrong direction. So, if the cat cannot leap to the ground, it must descend hind feet first, basically sliding down the branch while it alternates its limbs in disengaging the claws from the bark.

What makes margays so acrobatic are flexible hind legs, especially the toes, and ankles that can rotate 180 degrees. Also known as tree ocelots for their arboreal agility, margays can grasp and cling to a limb with their hind feet as well as they can with their front feet, enabling them to race down trees as well as up them. Margays can also "run" along the undersides of branches. Clouded leopards, reputed to be equally agile, are even able to hang from branches with their hind feet.

HOW DO CATS SEE?

A noticeable feature of the cat skull is large eye sockets placed well forward on the face. This tells us two things. First, large sockets indicate that the eyes themselves are large, suggesting that cats have good vision. Loosely speaking, large eyes may also gather more light than small eyes, so their relatively large eyes give cats an advantage in low-light conditions. Second, the placement of the eyes gives cats

Climbing up a tree is easy for most cats, even those whose natural habitat is largely treeless, such as snow leopards. Cats climb to escape from predators and to rest out of sight; only a few species actually hunt in trees.

binocular vision, which is the ability to focus both eyes at once on a single object. Binocular vision allows cats to judge distances, which is important to pouncing accurately on their prey. Primates, other carnivorans, and predatory birds such as eagles and owls also have binocular vision. In contrast, the eyes of cows and other hoofed mammals and most birds are placed on the sides of the face. These animals cannot see the same object with both eyes at once; however, they can detect predators coming from two different directions without turning their heads. Among the Carnivora, cats have the greatest degree of binocular vision, although it is less than that of primates. However, this means that cats have a wider visual field than ours—295 versus 210 degrees—giving them better peripheral vision than we have.

Eurasian lynx, like most cats, hunt primarily by sight rather than sound or smell, but they are reputed to possess particularly good vision among the cats. In one experiment, a captive lynx could see a mouse at 75 meters, a hare at 300 meters, and a roe deer at 500 meters. The origins of the word *lynx* reflect the belief in these

The pupils of cats change in size according to light conditions. They become round and large, as this puma's are, to let in as much light as possible when it is dark and contract to keep from being dazzled in the daylight.

The pupils of ocelots and many small cats contract to narrow slits in bright light, letting in almost no light, while those of big cats remain oval. This may reflect the more nocturnal behavior of small cats, which are less likely to be active during the day than large ones.

animals' visual acuity. From a Greek word meaning "to shine," *lynx* is signified in the name of the mythological Greek hero Lynceus, who was able to see through stone and see things underground.

HOW CAN CATS SEE IN THE DARK?

Evidence from domestic cats shows that they can see in light one-sixth as bright as what humans require. And because most cats hunt at night or at dawn and dusk, we assume that other cat species possess similarly good night vision. What makes this possible?

We mentioned that a cat's large eyes in general may gather more light. However, it is the size of the pupil that actually determines the amount of light that enters the eye and hits the light-sensitive area at the back of the eye, called the retina. You might think, then, that the bigger the pupil, the better for hunting and other activity in the dark. But the pupil can't be so big that the cat is blinded by light, because most cats are also active at times during the day. Among people, this problem is solved by the pupils' response to changing light levels: they enlarge in low-light conditions and contract in bright light. (This is why we are briefly blinded if we go directly from a dark movie theater to the sunny sidewalk.) Cats' pupils do the same thing, but to a greater degree, growing very large and very small depending on the ambient light. Relatives of the domestic cat even have a unique muscular arrangement that closes the pupil almost completely to two tiny pinholes at either end of a vertical slit. Big cats have elliptical pupils, which are round when dilated in the dark but are long and narrow in the light.

Cats' larger pupils are associated with other differences in their eyes as well. With increased pupil size, the lens must also be larger to prevent distortion at the edge, and it and the cornea must be more strongly curved so that the light is focused most closely on the retina. The more strongly curved cornea creates a larger anterior chamber inside the eye. These changes are more pronounced the more nocturnal the animal is. Lynx hunt mostly at night and have larger anterior chambers, lenses, and pupils than people have. Pumas, which hunt day or night, are intermediate.

Another adaptation for night vision is a structure called the tapetum lucidum, which lies behind the retina. Like a mirror, the tapetum lucidum reflects the light that hits it back through the retina to produce a brighter image. Basically, the same light gets used twice. This is what causes the "eyeshine" you see when you catch a cat (or many other animals with this visual structure) in your headlights at night.

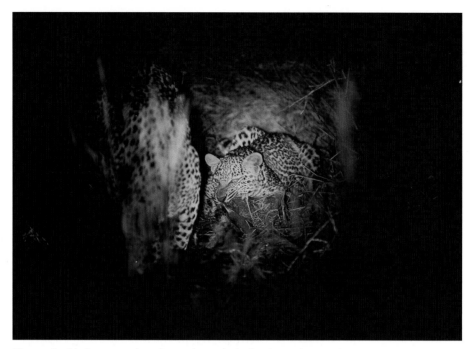

The artificial lights we need to observe leopards feeding hungrily at night give an eerie, other-worldly air to these animals' natural nocturnal behavior. Cats need about one-sixth the light we do to see.

ARE CATS COLOR-BLIND?

Strictly speaking, cats are not color-blind, but their perception of color is limited. Two kinds of light-sensitive cells—cones and rods—are found in the retinas of cats, people, and other mammals. Cones detect bright light, and the pigments they possess provide the ability to see color when hit by light of different wavelengths. Rods activate under low light and lack the pigments that produce color vision. As you might expect, rods dominate the retinas of cats, although an area of cones occurs in the center of the retina. Some of these cones are sensitive to green light and some to blue, but none to red, so cats are limited to what is called dichromatic vision. These cones let cats see over a narrow range of wavelengths during the day. Nonetheless, cats do not seem to pay attention to colors, and only with difficulty can they be trained to discriminate colors. Color is probably just not relevant to cats; in the dark of night, a red animal is no easier to catch and kill than a blue one.

In evolving the ability to see well in the dark, cats sacrificed color vision; they also sacrificed a certain amount of visual acuity. Splotches of dappled sun and shade playing on the forest floor—to us, a beautiful setting for a spotted cat—look like only a gray blur to the cat itself.

In a remarkable experiment, neuroscientists from the University of California at Berkeley were able actually to see the world, or at least a small part of it, through the eyes of a domestic cat. The scientists hooked up electrodes to 177 cells in the cat's thalamus, a part of the brain connected to the retina through the optic nerve. The electrodes recorded the cells' response to visual stimulation and relayed this information to a computer. As is well known, different brain cells fire most readily to different stimuli: some to edges, some to lines of various angles, and so on. Using a highly sophisticated analysis, the scientists transformed the responses of the cat's brain cells and videotaped them to produce images of what the cat saw. The results were astonishing: on the videotape the scientists were able to see the cat's perception of the images they had shown it, including a human face and a natural scene with trees (Stanley et al. 1999).

HOW DO CATS HEAR?

One reason our pet cats sometimes seem clairvoyant, or their behavior seems inexplicable, is that they can hear things we can't. Domestic cats have long been subjects in studies of hearing, and the similarity among cats makes it reasonable to extend the findings to their wild relatives, even though there is some variation. A cat's hearing is far more sensitive to high-pitched sounds than a human's is. While the upper limit of our hearing is about 20 kHz, a cat's is about two octaves higher, or up to 60 kHz. (In fact, measuring responsiveness of auditory nerves reveals that cats can hear tones up to 100 kHz, but such tones have to be so unnaturally intense, or loud, that this sensitivity is not useful to cats.) Cats' sensitivity to high-pitched sounds enables them to detect the squeaks and chirps of rodent prey, which use ultrasounds in the range of 20 to 50 kHz to communicate.

However, cats need more than the extrasensitivity of their auditory nerves to hear rodent calls, which are not only high but also very quiet. Relatively large external ears (pinnae) with additional surface areas, created by convolutions on the inside of the pinnae to collect the sound waves, give cats yet more hearing assistance. What's more, the pinnae are highly mobile, each controlled in the domestic cat by 30 muscles (compared with 6 such muscles in humans' outer ears), giving them mobility to better find the sounds. Servals, which hunt rodents scurrying through the tall grass of African savannas, possess relatively the largest external ears of any cat, enabling them to detect their prey by sound rather than by sight. Servals can even hear rodents moving through underground burrows, so they can nab these prey when they emerge. The Eurasian lynx reportedly can hear a hare nibbling 50 to 60 meters away.

All cats have relatively large, mobile external ears that enhance their ability to detect the sounds of potential predators or prey. The especially large ears of the caracal suggest that its hearing is particularly acute at some frequencies, as is the hearing of some other cats that live in arid habitats.

As cat species get bigger, however, auditory sensitivity shifts toward lower frequencies. In general, larger animals both hear and produce lower-frequency sounds than smaller animals. In a thorough comparative study of the middle ear in a variety of cat species, G. T. Huang and his colleagues found that the middle ear's size and sensitivity are scaled to body size, with low-frequency sensitivity increasing with size (Huang et al. 2000).

Among cats, larger species hunt larger prey, so, in accordance with this rule, the sounds their prey make will be deeper. Further, large cats, such as lions and tigers, use sounds to communicate at a distance, and low-frequency sounds carry farther. The rumble of a lion's roar may be below 0.2 kHz and as loud as 114 decibels. Roars can be heard by other lions, and by people, up to 5 miles away. Recently, scientists determined that tigers are most responsive to sounds around 0.5 kHz but are capable of producing very-low-frequency infrasonic sounds at and below 0.018 kHz! Presumably, they can also hear these sounds, although this has not been demonstrated in any studies. Domestic cats, however, despite their small size, have good hearing for both low (0.2 kHz) and high (60 kHz) frequencies, so it would not be surprising if tigers and lions could hear even lower frequencies.

Huang found an interesting deviation from the correlation between middle ear size and sensitivity and body size. Sand cats are about the size of small domestic cats but are far more sensitive to frequencies below 2 kHz than domestic cats and similarly sized wild cats are (Huang et al. 2002). Both the external and middle ears also have unusual structural features, and the auditory bullae are unusually large (see the discussion of these in *What Are Cats?*). Huang believes this is related to how sound is propagated in the arid desert environments in which sand cats live. In dry environments, the air absorbs high-frequency sounds more readily than low-frequency ones, meaning that low-frequency sounds can be heard from farther away. Interestingly, sand cats' long-distance call is quite loud, more of a bark that a meow, and has strong components at low frequencies. Further, cats living in dry, open habitats less severe than the desert, such as the black-footed cat and Pallas' cat, have larger than predicted bullae, but not so large as those of the sand cat.

The ability to sense vibration is related to both hearing and touch. Sound is composed of cycled air vibrations that hit the eardrums in a particular way, whereas other vibrations are detected by receptors deep in the skin, although they work in a similar fashion. Cats are said to be quite sensitive to substrate, or ground, vibrations.

HOW DO CATS TASTE AND SMELL?

Taste and smell together are called the chemical senses. A little is known about the sense of taste in domestic cats, and nothing is known about this sense in the other species, but it does not seem terribly important, because cats possess only about 500 taste buds compared with our 9,000. Domestic cats perceive these tastes, in order of importance: sour, bitter, salt, and, least of all, sweet. Taste buds are absent from the center of the tongue, where the papillae are modified to be strong, backward-pointing, and rough, like a rasp, and serve to scrape bits of meat off bones.

Cats use two organs to smell, or, more broadly, for olfaction. One includes the chemical receptors on the interior lining of the nostrils, which are what we think of when we talk about smell. The other is called the Jacobsen's, or vomeronasal, organ, consisting of two tiny openings on the roof of the mouth through which chemicals send messages to parts of the brain concerned with sexual behavior. Male cats perform flehmen, a behavior in which they curl their lip into a grimace after sniffing a female to detect whether she is in estrus (receptive to mating). Flehmen may serve to draw chemicals into the vomeronasal organ. (Whether this organ exists and is functional in people remains controversial.)

Cats' acute sense of smell—not as good as dogs' but 30 times betters than ours—enables these lions to detect and follow a trail of blood across the savanna. Such a trail may lead to a carcass they can steal from some other predator.

The olfactory mucosa that lines the nasal cavity in cats contains some 200 million special olfactory receptors, and the cat's mucosa is twice as big as that of humans. Cats are estimated to have a sense of smell 30 times better than ours.

Unlike many other carnivores, cats do not generally use smell in hunting, and their olfactory abilities are not as acute. This is reflected in the size of the brain area devoted to olfaction. In dogs, for instance, the olfactory lobes form about 5 percent of the brain volume, whereas in cats they form less than 3 percent. Olfaction is important in the social and sexual life of cats, however, and most communication between cats occurs through scent marks (see *How Do Solitary Cats Communicate?*).

HOW DO CATS SENSE THE WORLD THROUGH TOUCH?

Overall, a cat's skin is quite sensitive to touch. Specialized sensory hairs occur all over a cat's body, but, except for the whiskers, little is known about how these are

Touch-sensitive facial whiskers, called vibrissae, are like fingers groping in the dark, helping cats to maneuver successfully even in total darkness.

used. Feline "whiskers," called vibrissae, are coarse, stiff hairs that basically extend the sense of touch beyond the surface of the skin. Richly endowed with nerves, vibrissae detect minute variations in air currents caused by objects in the environment. Vibrissae are found in three groups in cats: on the cheeks, above the eyes, and on the muzzle. Touch-sensitive vibrissae enhance a cat's ability to move accurately, especially in the dark. Cheek vibrissae reveal this function. A resting cat holds its whiskers straight out at the sides, a sniffing cat holds them back against its cheeks, but a walking cat fans them forward, in the direction in which it is moving. Whiskers are also fanned forward as a cat pounces on prey, helping the cat to judge exactly where to deliver a killing bite. With input from whiskers, a blindfolded domestic cat can even orient itself to kill a mouse with a neck bite. Cats also have vibrissae on their wrist, or carpal, joints to assist them in placing their feet properly without looking when stalking. Areas of the domestic cat's brain devoted to processing visual and tactile signals are similarly mapped and adjacent to one another.

This suggests that sight and touch act in a coordinated way to enable a cat to navigate a complex world.

WHAT ARE CATS' COATS LIKE IN TEXTURE, AND WHY ARE THEY STRIPED, SPOTTED, OR PLAIN?

All cats wear beautiful fur coats. Fur protects the skin, enhances the sense of touch, and insulates the body from temperature changes. Species living in the coldest climates, such as Eurasian and Canada lynxes and snow leopards, have the longest, most densely furred coats, although there is seasonal variation. A snow leopard's winter coat can be more than 5 centimeters long on the back, sides, and tail, and nearly 12 centimeters long on the belly.

Across its large range, the leopard's fur also varies in texture and length in a way that may be related to climate. African leopards (*Panthera pardus pardus*) have short, coarse fur, while Amur leopards (*P. p. orientalis*) have long thick fur to keep them warm in the cold winters of northeastern Asia.

The color and markings of each species are unique, but there is wide variation within each species as well. The most likely scenario for the evolution of fur patterns suggests that the original, or ancestral, forms had flecks or small spots and that other patterns have derived from this. For instance, the rosettes of jaguars are small spots organized into patterns, while stripes are flecks merged together.

Scientists believe that a cat's coat color and markings are primarily designed for camouflage, although they may also play a role in thermoregulation and communication. A cat must conceal itself while stalking prey, and cats of most species must also conceal themselves from animals that prey on them. So markings of each species tend to blend into that cat's own natural habitat, making it nearly invisible there, even though the markings of some species, such as the bold stripes of tigers, seem conspicuous when we see the animals in photographs or in a zoo.

The background color of the fur generally corresponds to the predominant colors of the cat's natural landscape. Depending on the uniformity of the landscape, the fur may be plain or lightly marked, or it may be striped, spotted, or blotchy to break up the outline of the body. The plain, pale fur of a sand cat matches the color of desert sand; the similarly plain fur of a Pallas' cat lets it disappear among barren rocky outcrops; and the tawny coat of a lion blends into the dusty savanna.

In contrast, more boldly marked cats tend to be denizens of forests and woodlands, and many are all or partly arboreal. Spotted and blotched cats, from jaguars and ocelots of South America to clouded leopards and marbled cats of Asia, are

(left, top) **In another step or two, this tiger's black-and-orange striped coat, momentarily conspicuous in the clearing, will help it disappear among the vegetation. Camouflage coloration enables cats both to approach prey undetected and to conceal themselves from predators.**

(left, center) **Until it can begin to outrun predators at about four months of age, a baby cheetah wears a long gray woolly coat over the spots on its back and belly. Baby cheetahs are very frequent victims of predators, such as lions and hyenas, and this coat may offer better camouflage to hiding cubs than the more darkly spotted fur of adults.**

(left, bottom) **Why the tails of many cats are tipped in black or in white, as this leopard's is, is not known. Scientists speculate it may be important in communication or it may distract predators from attacking the body.**

(top, right) **Bold spots, blotches, and stripes are typical of cats that live in the variegated vegetation of forests, whereas these lions and other cats living in landscapes of more uniform coloration tend to have tawny, plain coats.**

forest residents, and their fur may resemble dappled light filtering through leafy vegetation. Tigers, with their striking black stripes on an orange background, also become nearly invisible in such dappled light, and their vertical stripes match the tall grass in which many of them hunt.

The young of some plain-furred cats, including lions and pumas, are born with spots. Scientists speculate that spotted fur offers better concealment to the helpless young in their nests of brushy vegetation. However, spotted fur is considered to be the ancestral pattern in cats, so spotted young may simply be a relict of evolution. The spots gradually disappear as the cubs mature.

Species of cats with very large geographical ranges that may encompass many diverse habitats often display considerable variation in coat colors and patterns. Wildcats (*Felis silvestris*), the ancestors of domestic cats, live throughout Eurasia and Africa; their markings are heavier and darker where they live in forests and are less pronounced and lighter where they live in open savanna habitat.

On the other hand, jaguarundis exist in two color forms in the same areas, which once caused some observers to label them as two separate species. The plain fur of the jaguarundi falls into two color categories: reddish and gray, with a range of variation within each category. Thus a "gray" jaguarundi may be nearly black, and a "reddish" one may be tawny (see *Why Are Some Cats Black?*). The coats of the Asian golden cat are generally plain, ranging in color from golden to red to gray, but individuals with spots and stripes are also seen.

In many species, even plain ones, distinctive facial markings are apparent, for example, the black "teardrops" of cheetahs and the white eye rings seen in ocelots, margays, lynxes, and others. The functions of these facial patterns are unknown. Some speculate that the black teardrops of cheetahs reduce glare (some football players paint black lines under their eyes for this reason), but this has not been demonstrated in any studies. Many cats also have black-tipped tails, again for unknown reasons, but ideas about their functions include their use in communication, in enabling young to follow their mothers, or in either attracting the attention of prey or deflecting the attack of a predator away from the cat's body.

WHY DO CATS HAVE WHITE SPOTS ON THE BACKS OF THEIR EARS?

In about half of all cat species, most of them forest-dwellers, prominent white spots or bars adorn the backs of their ears. Because these spots are visible in even very low light, scientists speculate that they may act as beacons that help young cats to trail their mothers through the forest or tall grass.

The prominent white bars that adorn the backs of the ears of about half of the cat species, including tigers (top), may help youngsters follow their mothers along dark forest trails. Most of the species with these white ear markings live in forests, making grassland-living servals (below) exceptional in this regard.

WHY ARE SOME CATS BLACK?

Everyone has heard of a black panther, which in reality is just a leopard with black background fur, which obscures the cat's spots. It is not a separate species; in fact, black-furred and spotted cubs can be siblings born in the same litter. Known as melanism, black fur in leopards and domestic cats is the result of a single recessive gene for coat color, whereas in jaguars it results from a single dominant gene. Melanistic, or black, individuals appear in many, if not all, species but do so frequently among leopards, jaguars, margays, ocelots, kodkod, oncilla, and Geoffroy's cats, less often among wildcats, caracals, bobcats, pumas, servals, cheetahs, clouded leopards, and lions. Black tigers have been reported but may be extremely rare.

(above) As can be seen in this black, or melanistic, Geoffroy's cat, dark background fur conceals the spots that more typically are visible on lighter background fur in this cat, in leopards, and in other species in which black individuals sometimes occur.

(right) An intriguing new hypothesis suggests that the genes responsible for the black fur of some leopards may be associated with increased resistance to viral infection. This may explain the prevalence of this mutation in leopards living in rain forests compared with leopards living in other habitats, where black fur is rare.

Gloger's rule suggests that dark fur is correlated with living in humid, rain forest habitats, although the reasons for this are unclear. Among leopards, black individuals are very common in tropical Asian rain forests, where dark color may be an advantage in the murky dim light that reaches the forest floor (up to half the leopards in Java and nearly all in Malaysia are black), but black individuals are rare elsewhere. Black forms of both leopards and servals are found in the Aberdare Mountains of Kenya, which are covered in fairly humid forest, but are rare in the African savanna habitats. Dark forms of jaguarundis are more common in rain forest habitats, whereas the reddish ones occur more often in dry habitats. Similarly, tigers living in hot Sumatran rain forests tend to be darker than tigers living in colder climates, although the rain forest individuals are not black.

Recently, a novel hypothesis has emerged: genes for blackness, or darkness, may provide resistance to viral infections! New genetic analysis has revealed that the gene that makes domestic cats black is different from the one that makes jaguars black and jaguarundis dark, and that both of these genes are different from the genes responsible for melanism in five other species. Moreover, although the same gene makes jaguars black and jaguarundis dark, the mutation in that gene differs between the two species. This suggests that the mutations involved in making cats black or dark have evolved at least four times and perhaps as many as nine times, and we know that any trait having multiple origins and persisting over time in related species is likely to be adaptive. The adaptive explanation is further supported by genetic analysis showing that the dark jaguarundis are, in fact, the mutant form, and they appear to be replacing the ancestral red forms. The jaguar and jaguarundi gene linked to melanism is similar to other genes that code for certain receptors on cell membranes—receptors through which, in humans, viruses such as HIV sneak into cells and take them over. However, some mutations make these receptors more resistant to invasion and therefore confer some immunity to HIV and related viruses. Thus, scientists speculate, mutations associated with melanism may give cats resistance to certain viruses.

It could be that humid rain forest habitats harbor such viruses—a possible explanation for Gloger's rule. Or, in the case of leopards, a viral epidemic in the past may explain the large number of black animals in Java and Malaysia, because those with a mutation that leads to both melanism and resistance would survive better than those without it. Something similar may account for the Aberdares as a hot spot for melanistic leopards and servals in Africa.

WHY ARE SOME CATS WHITE?

The most famous white cats are white tigers, seen in some zoos and circuses and extremely rarely in the wild. It is a common misconception that white tigers are Amur (Siberian) tigers (*Panthera tigris altaica*), whose coat color is adapted to blend into snowy landscapes. Amur tigers do tend to be paler than tigers elsewhere, but they are not white. On the other hand, neither are white tigers true albinos, which lack all pigmentation. Albinos are very rare among cats and usually are deaf, blind, or have other serious, usually fatal, problems. Rather, white tigers are intermediate in pigmentation, sometimes called leucism (*leuc-* means "white"), because of a mutation that, in domestic cats, occurs at a gene locus with three variants: chinchilla, acromelanic (as in Siamese cats), and complete albino. White tigers have white background fur with dark brown stripes. White is a recessive mutation of the gene for normal orange color and is associated with reduced pigmentation in the eyes as well, which are blue (complete albinos have pink eyes), as are the eyes of Siamese domestic cats, which similarly have recessive genes for a pale coat color.

While many people consider them particularly beautiful, white tigers usually suffer from visual and other abnormalities. White tigers are extraordinarily rare in the wild, and those in US zoos and circuses are products of artificial selection. As a result, breeding white tigers does nothing to help save wild tiger populations.

White or very pale tigers appear naturally from time to time in India, but most white tigers in zoos and circuses today derive from a single male white cub captured in India in 1951. He was bred with a normal-colored female, and their young were normal-colored too. But then he was mated to one of his daughters, and this union produced four white cubs. Breeding white brothers and sisters yielded more white cubs, revealing that whiteness is the result of an individual inheriting two copies of a single recessive gene. One of the cubs, a female, was sent to the Smithsonian's National Zoological Park in Washington, DC, where she founded a line of white tigers in the United States, which has been perpetuated by some zoos through selective breeding. Only a few responsible, conservation-oriented zoos maintain or breed white tigers today, because they are of no value in conservation breeding programs and take up space that might otherwise be occupied by other big cats. Some people claim that breeding white tigers is an effort to save an endangered race because they are so rare in the wild. However, they are naturally very rare in the wild, and there is no reason to maintain their recessive genes artificially.

Most white tigers suffer from visual abnormalities that are also found in Siamese cats. The visual pathways in their brain are wired differently, so that some of the optic nerve fibers actually go to the wrong side of the brain. This leads to cross-eye and a squint in some white tigers and Siamese cats, although even individuals without obvious cross-eye or a squint have the same abnormal visual system.

Although there had long been rumors of white lions in South Africa, the first ones were captured in 1975 near Kruger National Park, which seems to be the only place this mutant form has appeared. These lions were bred to form a line of white lions, as were others captured later by the Johannesburg Zoo in South Africa. Today there are white lions in several zoos, and they have been bred, along with white tigers, by the Las Vegas act Siegfried and Roy. Like white tigers, white lions appear to result from a single recessive gene. They tend, however, to have normally colored eyes, with only a few showing the blue eyes typical of white tigers.

DO MALE AND FEMALE CATS LOOK ALIKE?

Apart from the expected differences in external genitalia, male and female cats of most species look alike, but males are usually larger than females. Size dimorphism may be extreme, especially in big cats such as lions and leopards, where males may be 50 percent heavier than females. Among smaller cats, males tend to be only 20

In most species, male and female cats look alike, but males are usually bigger and more robust than females, which is clearly evident in this image of caracals. The male, on the left, is husky and has a larger head than the slighter female.

Unique among cats, the adult male lion's lavish mane makes it easy to tell male and female apart, but size is often an equally obvious difference between the sexes. Males in big cats, such as lions, may be 50 percent larger than females. Intense competition for females puts an evolutionary premium on being big, and big males' competitive edge may make them more attractive to females.

percent heavier than females (see *Why Are Male Cats Larger Than Females?*). Moreover, male cats have relatively larger, broader heads, larger teeth, and greater biting force than females. Lions are unusual in that the two sexes look obviously different: many males possess manes of long fur, but no females do. To a less obvious degree, this is also true of tigers: males have a pronounced sagittal crest on their head and a ruff around the face. Also, male and female leopards have differently shaped skulls, suggesting they have different food habits.

WHY ARE MALE CATS LARGER THAN FEMALES?

It is not a given among mammals that males will be larger than females; neither are males larger than females in all other species of predators. Females in most birds of prey species are larger than males. Among the mammals, 25 percent of the families have at least one species in which the females are larger than the males. In the Carnivora, females of mongooses and hyenas are larger than males.

Some cat species show great size dimorphism, with males larger than females, and some species show little or none. Also, size dimorphism between the sexes in some species varies across their range; adult male leopards can be as little as 30 to as much as 50 percent heavier than females. Moreover, the largest males sometimes weigh three times as much as the smallest males.

Biologists think that, in many animal species, larger body size in males has evolved because of the advantage of large size in winning male-to-male combat, thus increasing opportunities for mating. Aside from these contest results, females may choose larger males to breed with also because their greater strength makes them seem more fit. This is called the sexual selection hypothesis. Some extreme examples occur among mammals in which males fight for harems, such as in some larger ungulates and seals, and for prides, as in lions.

But what sets the body size of females? Some think the body size of the female is determined by what is optimal for the species under prevailing ecological conditions. Katherine Ralls of the Smithsonian's National Zoological Park argues that the degree of sexual dimorphism in size is the result of the difference between the sum of all the selection pressures affecting the size of females and the sum of all those factors affecting the size of males (Ralls 1978). So males may be sized to win contests to gain access to reproductive females, whereas females may be sized according to the size of their young and the energetic demands of pregnancy and lactation.

The rate at which animals use energy depends on their body weight (see *What Does Size Have to Do with It?*). The prevailing ecological conditions have an influence here. When and where food is ample, larger females may be favored, but if food is scarce, smaller females may have an advantage, because they require less food in total to maintain themselves and, among mammals, are able to put more energy into milk production. Large body size is an advantage if females can store large amounts of energy as fat deposits during periods of superabundant food availability, to be used during lean times. However, wild cats don't work this way; they must have a constant supply of prey year around. Therefore, small female cats, especially those living in unstable or seasonal environments, may have the advantage, because they need absolutely less food but also can lay down some fat stores rapidly just before reproduction.

Sexual difference in body size or morphology may also evolve to adapt the sexes to significantly different ecological niches. Individuals of one sex compete with individuals of the other sex when resources are limited—as they usually are for cats—just as they compete with individuals of other species. The first is an example of intraspecific competition; the second, of interspecific competition. Intraspecific competition results in selective pressure to increase sexual dimorphism, which then decreases that competition. However, competition and food availability may also intervene to constrain the range of feasible body sizes of each sex. Sexual dimorphism is most pronounced in the absence of interspecific competition. For instance, tigers are absent from the island of Sri Lanka, so here male leopards are larger than those that share the tiger's range in India; without interspecific competition, intraspecific competition increases among the leopards, creating greater size disparity between the sexes in this part of their range.

WHAT IS THE MAIN PURPOSE OF MANES IN LIONS?

Three related functions have been ascribed to the manes of male lions. First, the mane may protect a male's head and neck during combat with other males. Second, it may be a visual signal that makes a male look bigger and thus intimidates his rivals. Finally, the mane may be attractive to females.

In general, the manes of the Asiatic subspecies of lion are sparser than those of the African subspecies, but recently a population of lions living in Kenya's Tsavo National Park was found in which males mostly lack any mane, and none have the long, flowing manes seen in other populations. Two other now-extinct subspecies, the Cape lion and the Barbary lion, had magnificent long dark manes that extended to cover their belly. Male lions living in cooler climates have longer, darker manes than those living in hotter areas.

Male lions begin to grow manes at about 1 year of age, and manes reach full size at about 4 years of age. Mane growth and maintenance is controlled by the male hormone testosterone; a castrated male quickly loses his mane. However, the mane of an individual male may change as often as monthly or from year to year. Manes vary from sparse blond fur around the face and neck to flowing dark tresses up to a foot long. In the most heavily maned lions, long thick dark fur covers the forehead, neck, shoulders, throat, and belly; tufts of fur may also spring from the elbows. In contrast, a very lightly maned male, such as those seen in Tsavo, has short sideburns, a short crest along the back of the neck, and a tuft on the chest.

Despite years of speculation about the mane's function and the significance of the variation, only recently were studies conducted to attempt to answer the questions. University of Minnesota graduate student Peyton West and her adviser, Craig Packer, who has studied lions for more than 20 years, carried out this work with fascinating results (West and Packer 2002).

First, West and Packer discarded the idea that the mane protects a male's head and neck—male lions do not attack this area more than any other, and wounds there are no worse than wounds suffered elsewhere—and focused on the mane's role in male-to-male competition and in attracting females.

A male lion's mane says a lot about him. Males with long, dark manes are older, healthier, and more aggressive than males with short, light-colored manes. Dark-maned males attract females and intimidate other males, and females spurn light-maned males, which are more often picked on by male rivals.

West set up two lifelike lion dummies and adorned them with contrasting manes—long versus short, and dark versus light—then watched to see how both male and female lions responded to them. Playbacks of hyena calls were used to attract live lions to the dummies, because lions often chase hyenas off carcasses and take the food for themselves. For females, color was the key; they preferentially approached the dark-maned dummies and were seemingly indifferent to mane length. Further, long-term data collected by Packer and his students revealed that females most often mated with the male in their coalition with the darkest mane, again ignoring its length.

Males, on the other hand, always approached the light-maned dummies when choosing between those and the dark-maned and more often approached the short-maned dummies when choosing between those and the long-maned. West and Packer's further analysis showed that males with dark manes were older and had higher levels of testosterone, which supports higher levels of aggression; older males also had longer manes. Short manes signaled relative youth and were also associated with males having a serious injury. In one of their study sites, the Ngorongoro Crater, where food availability didn't change during the year, dark-maned males also ate more than light males.

Female lions have a lot to gain by selecting mature, aggressive, healthy, well-fed males as mates. Such males are most likely to withstand the attempts of other males to evict them from a pride, and there's nothing female lions like so much as stability. New males invariably kill all the infants present when they take over, so the longer one group of males hangs around protecting females and young, the better. And, in fact, West and Packer found that coalitions with dark-maned males lasted longer, and their young were less often wounded. Clearly, male lions read the same message in the manes, because they avoid males whose dark manes show them to be capable of putting up a good fight but risk a confrontation with a light-maned male that, thanks to youth or infirmity, may be easily bested.

WHAT ARE THE BIGGEST CAT AND THE SMALLEST CAT?

Several species run neck and neck for the title of smallest cat. Two small species live in Asia—the rusty-spotted cat of Sri Lanka and India, in which adults may weigh only about 1 kilogram and be 35 to 48 centimeters in head and body length, and the flat-headed cat of Malaysia, Thailand, Borneo, and Sumatra, at 1.5 to 2.2 kilograms. In southern Africa, among the black-footed cats, females weigh about 1.5 kilograms, and males, from 2 to 2.5 kilograms; head and body length is 36 to 45 centimeters. Sand cats, also in Africa, range from 1.3 to 3.4 kilograms. In South America, both kodkod and oncillas are reported to weigh from 1.5 to 2.8 kilograms. Among domestic breeds, the Singapura is generally credited as the smallest, with females in the range of 2.3 to 2.7 kilograms and males from 2.3 to 3.6 kilograms.

One difficulty in awarding the title of smallest cat is that so few individuals of some small cat species have been reliably weighed and measured. Another is that cats of an individual species may vary in size across their range. Leopard cats (*Prionailurus bengalensis*), which are widely distributed in Asia, are generally credited as weighing between 3 and 7 kilograms, with head and body length between 44 and 107 centimeters. But a recent study of leopard cats in Malaysian Borneo reported male weights at 1.8 to 2.9 kilograms and head and body measurements of 43 to 50 centimeters, with figures for females at 1.7 to 2.3 kilograms and 43 to 47 centimeters.

The biggest cats by far are tigers and lions, with tigers generally getting the nod. A very big male tiger living in India or Nepal, where the largest tigers now occur,

"Microcats," such as this rusty-spotted cat (left), native to India and Sri Lanka, weigh from 1 to about 3 kilograms. These tiny predators often have huge appetites, however, and hunting is an all-consuming task. In a size class all their own, lions (right) and tigers vie for the title of world's largest cats. On average, cats in the next size class, such as leopards and jaguars, weight about one-fourth as much, with the smallest female lion usually outweighing the largest male leopards and jaguars.

may reach 295 kilograms and stretch as long as 250 centimeters. The largest male lion may tip the scales at 250 kilograms and measure 250 centimeters. On the other hand, any given individual of one of these species might weigh or measure more or less than an individual in the other species. In general, female tigers tend to be smaller than male lions. And tigers living on the island of Sumatra (*Panthera tigris sumatrae*) may be half the size of tigers from Nepal (*P. t. bengalensis*).

WHAT DOES SIZE HAVE TO DO WITH IT?

Size and Scale

Although weight varies among cat species by a factor of at least 300, this does not mean that enlarging an image of a domestic cat with your computer will result in its looking exactly like a big lion in all but its markings. As size increases, different parts of the body change either at the same or at different rates. Determining the size relationships between different features of an animal and how these relationships differ from taxon to taxon is the study of *allometry*, or scaling. The most basic of allometric relationships is the ratio of surface area to volume: as an animal gets bigger, its surface area increases (by squaring), but its volume increases much more rapidly (by cubing). Thus, a small animal has a relatively larger surface area–to–volume ratio, and a larger animal has a relatively smaller surface area–to–volume ratio. This means that a small mammal will lose heat faster than a large one, because heat is lost through its larger surface area and less heat is stored in its smaller internal volume. This is the basis for Bergmann's rule, which states that mammals tend to be larger in colder climates, although there are many exceptions.

The rate at which the size of one feature changes as the size of another feature changes can be described by the scaling factor, a single figure that is computed with well-worked-out mathematical formulas. When the scaling factor is 1, the two features are changing at the same rate over a range of body sizes; this is called *isomorphic allometry*. When the scaling factor is less than 1, one feature is changing more slowly than the other, which is called *negative allometry*; when it is more than 1, one feature is changing more rapidly than another, called *positive allometry*. Allometric relationships between body size and several other features are sufficiently well established to have achieved the status of laws. Kleiber's law, for instance, states that the basal metabolic rate in mammals decreases as size increases; the scaling factor for this is 0.75.

Many years ago comparative anatomist D. Dwight Davis looked at allometric relationships between body weight (mass) and the size of various body parts in several cats ranging in size from domestic breeds to lions. His most interesting results concerned the features of a cat that have the greatest effect on locomotion: the skeleton weight and the skeletal muscle weight. Physical laws dictate that it takes relatively more effort to move a large body than a small one; thus, for a lion to perform physical tasks as well as a domestic cat, its skeleton and muscle masses should show positive allometry. In fact, the relationship between body mass and muscle mass in lions has been determined as isomorphic, and the relationship between their body mass and skeletal mass is only very slightly positive, leading Davis to the conclusion that "the performance of a lion is certainly distinctly inferior to that of a domestic cat" (Davis 1962, 510). Other organs contributing to locomotor performance, such as the heart and respiratory system, also show either positive or isomorphic allometry in lions. However, their brain and eyeballs show considerable negative allometry; that is, as the size of the animal increases, the size of these organs decreases at a very fast rate. So, in your computer-enlarged image of the domestic cat, its eyes and brain would be much larger than a lion's, its heart smaller, and its skeleton and muscle about right.

Analyzing allometric relationships, especially those that deviate from the norm within a taxon, is often very revealing. For example, cheetahs have much larger hearts and lungs than a cat this size is expected to have, which relates to their ability to achieve more powerful bursts of speed than other cats.

If we return to the relationship between body mass and basal rate of metabolism, we see that mass may have the strongest influence on the basal rate in mammals, but differences related to diet, activity level, and climate remain. These account for most of the existing variation among the mammal species. Of the cats studied, all but two species have basal metabolic rates (BMRs) higher than predicted for carnivores of their size. Among carnivores, species that eat high-quality vertebrate prey, as cats do, tend to have relatively high BMRs, while species that eat low-quality food, such as fruits and invertebrates, tend to have relatively low BMRs. The exceptions among the cats are the arboreal margay and the jaguarundi, which, interestingly enough, include some fruits and invertebrates in their diet. However, arboreal carnivores also tend to be less active and have less muscle mass, so for these reasons margays may have relatively low BMRs even if they eat mostly vertebrate prey. Scientists could test this by examining the BMR of clouded leopards, a comparably arboreal cat. Some support came from a study of clouded leopards in zoos, which showed that these cats appear to have relatively low energy expenditures, conserving energy by minimizing their activity (Allen et al. 1995).

It is intriguing to note that black-footed cats eat voraciously and hunt almost continuously. Similarly, captive rusty-spotted cats have huge appetites, and wild rusty-spotted cats are very active, described as being like normal cats on speed. Sand cats, too, are very hearty eaters: a captive sand cat ate 15 mice in quick succession and may have eaten more had they been offered. All three species appear to eat whatever they can catch. These observations suggest that the smallest cats have relatively high BMRs, which require a lot of fuel, making them the hummingbirds of the cat family.

Individuals in about half of the 4,630 species of mammals weigh less than about 100 grams—the size of a small rat—making even the smallest of cats, at about 1,000 grams, large mammals. But within the cats, there is also much size variation, not only among species but also between males and females of the same species and among individuals in different parts of a species' range. For instance, pumas living in the tropics are half the size of temperate-zone pumas. Jaguars living in tropical rain forest are about half the size of those living in tropical floodplain savannas.

Fewer than 6,000 tigers, the largest cat species, remain in the wild, and an estimated 50 million feral domestic cats live in the United States alone, so being small appears to be advantageous, at least in some circumstances. For one thing, feral domestic cats, tend to produce two or three litters a year, and other small cats produce one or two; tigers can produce a litter only every other year. But most cat species are larger than domestic cats, so other circumstances clearly make being big a bonus. A cat's size, anatomy, behavior, and physiology are intimately related, and size is determined by several different selection pressures.

Size and Prey Selection

In the carnivores generally, and within the cats specifically, there is a general direct correlation between the size of the predator and the size of its prey, although this is not an absolute rule. Biologists use the ratio of prey size to female weight to make comparisons; a ratio of less than 1 means the cat is taking prey smaller than it is; a ratio above 1 means it is taking prey larger than it is. The average ratios are less than 0.1 for wildcats, servals, and caracals, 0.12 for Eurasian lynx, 0.3 for bobcats, 1.26 for snow leopards, 1.08 for leopards, 0.26 for cheetahs, 0.39 for jaguars, 0.52 for tigers, 1.06 for lions, and 2.44 for pumas. These ratios show that, on average, pumas take relatively the largest prey, and wildcats, servals, and caracals the smallest.

An increase in body size is possible only if a cat can increase its total food intake in an energetically feasible way. Chris Carbone of the Zoological Society of London and his colleagues, looking across the carnivores, found that at the 21.5 to 25 kilogram divide a major transition occurs in the size of prey that carnivores eat. Below this divide, carnivores prey on small animals that are less than half the weight of the predator; above this divide, the carnivores prey on large vertebrates as big as or bigger than the predator (Carbone et al. 1999). The smaller carnivores frequently eat invertebrates or a mix of invertebrates and small vertebrates, as the wildcat and the leopard cat do. Invertebrates and small vertebrates can be a superabundant food source, forming more than 90 percent of the total animal biomass in some areas. In spite of this abundance, larger carnivores cannot sustain themselves on such small prey; it is energetically too expensive to spend time pursuing these bite-sized items. Larger cats must take larger prey, even though there is less total biomass of larger vertebrates available, in part because larger vertebrates reproduce more slowly than smaller animals. However, larger cats usually have an advantage in their ability to take prey across a greater size range than smaller cats can.

Behavioral ecologist Louise Emmons (1987) studied medium-sized jaguars and small ocelots in the rain forests of South America. The two species together captured and ate the whole spectrum of mammalian prey, but they divided it into two size categories. Most (92 percent) of the ocelot's prey—35 mammalian species in all— weighed less than 1 kilogram; 85 percent of the jaguar's prey—also about 35 species— weighed more than 1 kilogram. The ocelot's small prey (mostly small rodents and opossums) reproduced multiple times in a year, had litters of three or more young, and females of these prey species could breed before they were a year old. The large prey, such as coatis, peccaries, and brocket deer, reproduced once a year, had litters of fewer than three, and females of these prey species could not breed before they were a year old. Ocelots ate 75 percent of the total of their prey available at any one time (what ecologists call the standing crop biomass), whereas jaguars ate only about 8 percent of the larger prey that was available at any one time. But because the smaller prey reproduced so much more rapidly than the larger prey, the ocelots' total take over one year was only about 6 percent of all the potential prey produced.

Emmons went on to compare all the wild cat species and found they could be divided into size categories on the basis of feeding tactics. The two great cats, lions and tigers, are about four times the size of the medium-sized cats—pumas, leopards, snow leopards, cheetahs, and jaguars—which are roughly two to four or more times the size of the small cats. For the great and medium-sized cats, Emmons found that the modal, or typical, prey size generally corresponds to cat size, although some take prey that averages about half their body size, and some take prey that is, on average, much larger than they are. When John compared the feeding habits of a leopard and a tiger in Nepal's Royal Chitwan National Park, he found that the female tiger killed prey that weighed four times more than prey killed by the female leopard living in the same area (Seidensticker 1976). Correspondingly, the leopard was about one-fourth the size of the tiger. For the small cats—those weighing less than 20 kilograms—Emmons's analysis revealed that the modal prey size is less than 10 percent of the cat's body size. (Among the lynx, which can be considered small cats, the modal prey size is greater than 10 percent, because lynx specialize in hunting and killing rabbits and hares.) Emmons's insight was a breakthrough, because she related the prey species and their reproductive potential to the size of their felid predators: the small cats of the world mostly eat rapidly reproducing prey, such as rodents and rabbits; the larger cats eat the slowly reproducing prey, larger ungulates (hoofed mammals), which produce only one or two young each year.

Because the smaller cats eat smaller prey, they have to catch prey—often many prey—every day. The bigger cats have to kill only one or two large prey in a week to live. Larger cats can travel farther each day, giving them a larger foraging radius than smaller cat species have; also, movement costs large cats relatively less energy than it costs small cats. The activity patterns of the different cat species is reflected in their prey in yet another way: prey of small cats is largely nocturnal, while that of the larger cats is an equal mix of nocturnal and diurnal. In addition, the small cats must compete with many other predators, such as raptors, snakes, and weasels, for their prey, and their prey species are also subject to population cycles, during which their numbers rise and fall sharply. Both of these aspects are less pronounced in the prey conditions of the large cats.

Size and Climate

Body size in some cat species may be related to climate. The races of warm-blooded animals living in colder climates are generally larger than the races living in warmer regions—this statement, known as Bergmann's rule, describes a general pattern that is seen in 72 percent of the bird species and 65 percent of the mammal species so far examined. We are most likely to see this pattern in cat species with fairly large geographical ranges, and we do see it in cheetahs, caracals, wildcats, servals, bobcats, jaguars, and pumas. However, leopards do not fit this pattern, nor do tigers, which both have very large geographical ranges. While the smallest tigers have occurred on the tropical Indonesian islands of Sumatra (*Panthera tigris sumatrae*), Java (*P. t. javanicus*), and Bali (*P. t. balica*) (the Bali and Javan tiger populations are now extinct), the largest tigers are from Assam in India and Nepal, and both are slightly larger than the most northern-living Amur (Siberian) tiger.

The explanation for Bergmann's rule is usually based on the allometric relationship in the ratio of surface area to volume: as an animal gets bigger, its surface area increases (by squaring), but its volume increases much more rapidly (by cubing) (see "Size and Scale," above). Heat production in homeotherms, or warm-blooded species, is related to volume, and heat loss is related to surface. Larger animals tend to produce more heat and lose less relative to smaller animals, giving them an advantage in cooler climates and disadvantaging them in hotter climates. Bergmann's rule also implies that smaller homeotherms, because of their relatively greater heat loss, have a problem obtaining enough food to meet their energy demands in colder climates. Many biologists have looked for the Bergmann pattern by examining body size in relation to latitude, assuming that the ambient, or surrounding, temperature increases with decreasing latitude. Others have found a better correlation between body size and a combination of ambient temperature and humidity, regardless of latitude, because humidity determines an animal's ability to unload heat by evaporation. Small body sizes are associated with areas of warm, humid weather; larger body sizes, with increasingly cool and dry regions. An animal will be able to unload heat more easily if it has a higher ratio of respiratory surface to body size, as smaller animals do.

In the more than 160 years since Bergmann proposed his rule, other hypotheses have been suggested to account for this phenomenon. For example, body size in some hoofed mammals may respond to latitudinal differences in plant growth, because larger body sizes are also positively correlated with abundant food resources. And larger body size may contribute to fasting endurance, an advantage in unpredictable environments. The larger the body, the more energy can be stored in the form of fat, and the longer an animal can go without food. Big northern raccoons, for example, grow very fat on superabundant fall foods, such as acorns, and then den-up and don't eat during the cold winter, when food is not available. Tropical-living raccoons don't experience such seasonal disparities in food abundance, and their resulting smaller size is an advantage in their climate.

Cats don't get fat, except in captivity. As specialized meat-eaters, wild cats never have a superabundance of food. However, larger cats can take larger prey and go longer between kills. In cold climates, where prey is preserved by the cold, a large prey such as a wild swine or deer provides many meals over many days. John once radio-tracked a young adult female puma that remained for three weeks near an elk she had killed (Seidensticker et al. 1973). In the wet tropics, such a large prey item lasts only a day or two before it rots. So in that climate, where a predator has to kill more frequently anyway, a cat has no apparent advantage to becoming any larger if it is already big enough to kill the largest prey available. In fact, there is a disadvantage to becoming larger, because in tropical environments, larger animals use more energy than smaller ones to keep cool.

The pumas John studied in Idaho's Salmon River Mountains, at about 45° north latitude, were very large, with males averaging 70 kilograms and females 45 kilograms (Seidensticker et al. 1973). Tropical pumas average half that size. Biologists at the University of Florida examined the size variation in pumas over their entire geographical range, from about 60° north latitude to 50° south latitude (Iriatre et al. 1990). The smallest pumas occur in the tropics, and the largest at the northern and southern extents of their range. The average size of puma prey also increases with latitude. A probable factor in pumas' smaller size at lower latitudes is their coexistence and competition with larger jaguars in the tropical lowlands, whereas jaguars do not live in the puma's range at higher latitudes. We do not know why the jaguar's range doesn't match that of the puma. Perhaps the powerful jaguars, with stocky, robust bodies built to catch and kill caimans (relatives of crocodiles) and giant turtles along with tapirs and deer about their own size, are simply outcompeted by lithe pumas in northern and southern mountain environments, where they kill elk and guanacos that are many times their own size. On the other hand, wolves and grizzly bears restrict how pumas use habitats in northern latitudes, and jaguars may once have been restricted by competition from now-extinct large carnivores, such as the giant short-faced bear. We do know that, within the tropics, jaguars that feed routinely on large prey are larger than jaguars that live in the rain forest and generally eat smaller prey.

Among tigers, the smaller ones live in the wet humid tropics; tigers in this environment can kill the largest ungulates available, but rain forests support few large ungulates, so tigers live there at a very low density, about 1 per 100 square kilometers. However, smaller tigers need less food and can live on the rain forest's more abundant smaller prey, such as small ungulates and primates. The biggest tigers on the Indian subcontinent live in habitats rich in large ungulate prey and reach densities of 10 or more adult tigers per 100 square kilometers. These highly productive habitats include moist tropical forests, forest edges, and tall grasslands of the savanna, where there is plenty of food for ungulates. These large tigers always live where they have access to water, in which they soak to cool off on hot, humid days. The large tigers at the northern edge of their range live at very low densities, as do large ungulates, which may exist at one-tenth the density they do in rich habitats in the Indian subcontinent. Smaller prey, including small ungulates and primates, available in more tropical environments, are not available to northern tigers. The northern environment is strongly seasonal, and the winters are very cold. When a large kill is made, it lasts long enough for the tiger to eat it all before it rots. Long fasting times between these kills is a way of life for the tigers living there.

Size and Interspecies Competition

Leopards do not comply with Bergmann's rule, but there is considerable variation in this cat's body size across its vast range, indicating that it is subject to differing selective pressures in different areas. In some parts of its range, the leopard is dominated by the larger lion and tiger, resulting in the leopard having to take smaller prey than it would if these other cats weren't monopolizing the larger prey. Across the range, leopards in populations that are limited to smaller prey are smaller than those in areas where larger animals can be taken as prey. Competition from one set of carnivores affects the body size of another set. Larger cats kill smaller cats when they catch them. Small cat species may actively avoid areas used by the larger cat species, thus channeling them into different habitats. More numerous large cats may reduce the amount of prey, especially large prey, available for the smaller predator. Bigger cats steal the prey of smaller cats when they can, a form of competition called *kleptoparasitism*. And cats have other competitors as well. Wolves kill pumas, and if the puma has a kill, a wolf pack will take it. Lions and spotted hyenas take kills from cheetahs and from leopards when they can, but leopards often haul their kills up into trees to store and eat them above the reach of lions and hyenas (see *How Many Different Cat Species Can Live in One Area?*). Thus, the larger cat species, through various forms of competition, can prevent the smaller cat species from getting bigger. Likewise, the smaller cat species and other smaller carnivores can keep the larger cat species from getting smaller, because they can outcompete the larger carnivore in exploiting small prey.

WHAT DO CATS EAT?

All cats are strict carnivores, eating mammal, bird, reptile, amphibian, and fish protein. Some of the smaller cats may include substantial amounts of invertebrates in their diets. Cats eat all or most of their prey, including meat (muscle), fat, bone, viscera, and other parts.

A cat must obtain enough energy to fuel growth, maintenance, and reproduction, or it dies. The small cats usually go only a few hours without food. If food is unavailable because of bad weather, for example, a cat may hole-up in a burrow or another protected site and rest to conserve energy. The large cats can go for days without food, but no cat can endure very long periods when food is unavailable. Cats require a regular source of animal food and cannot live in environments where significant seasonal food shortages occur. If their prey moves with the season, the cats move with it. Pumas follow their primary prey, mule deer and elk, from high-elevation summer ranges to lower-elevation winter ranges in the Rocky Mountains; cheetahs follow migratory herds of gazelles in the Serengeti.

Prey Size

What cats eat depends on the variety of prey available to them and the character-istics of that prey. Potential prey differ in the efficiency with which they can be found and killed. Other carnivores may compete for that prey. Environmental fac-tors and how vulnerable the cat is to other predators also influence their take. Be-cause cats live exclusively on other animals, they are already specialists. The opportunity to specialize further depends on the number and type of prey available and the cat's morphological and behavioral adaptations. Usually a compromise ex-ists between the number and type of prey available and the proficiency with which the cat exploits them. A jack-of-all-trades is a master of none, seems to be the rule. Our understanding of cat predatory behavior tells us that hunting success depends on the size of the prey relative to the size of the cat, and success may drop off sharply when prey is either too small or too large. From the cat's view, the young of an ungulate, for example, is different from the adult. Where John studied cats in Nepal, leopards were found to focus on hunting small ungulates of all ages, the young of medium-sized and large ungulates, and the adults of medium-sized ungu-lates; the larger tigers mostly killed adults of the medium-sized and large ungulate species (Seidensticker 1976). In other words, these cats were seeing prey not as an array of different species but as an array of different sizes.

Generalist or Specialist

Cats tend to specialize on prey of certain sizes, but most cats also live in environ-ments in which the prey they hunt and the conditions under which they hunt those prey vary both from one place to another and seasonally. Do cats exhibit flexibility in their hunting behavior when faced with such change? A specialist predator hunts in a stereotypic way and always exploits the same narrow range of resources in about the same way, but if it exploits this range of resources in various ways and includes new prey as they become available, it is said to be a flexible, or plastic, specialist. A generalist that hunts in a stereotypic way always exploits the same wide range of prey in about the same way; a plastic generalist exploits a wide range of prey by hunting in different ways.

Small rodents (rats and mice) are far and away the preferred prey of small cats. A single serval may catch and eat nearly 4,000 rodents in a year.

After rodents, birds are among small cats' favorite meals. Weighing in on the large end of the small-cat spectrum, caracals prey on large birds such as guinea fowl and francolins, which they may leap up to snatch out of the air.

Noted for their catholic tastes, leopards will not pass up a meal of hare, but they prefer larger prey, such as a variety of deer and antelope. Porcupines and monkeys are other major components of leopards' diets in some habitats.

Large hoofed mammals such as wildebeests and zebras are the mainstay of the diets of African lions. Groups of lions also tackle even larger ungulates, including buffalo, giraffes, and small elephants, and a single lion may make quick meal of a gazelle or hare.

Caracals, for example, weigh from 8 to 20 kilograms, and many people have the impression that they are bird specialists because of their impressive ability to leap and knock a bird out of the air as it explodes into flight from the ground. In a recent study conducted in the arid, coastal scrublands of South Africa, biologists N. Avenant and J. Nel (2002) found that caracals fed on prey ranging from 1-gram insects to 31-kilogram antelope, but their most common prey were rats and mice, which were also the most abundant prey. Predation on antelope was seasonal and limited. Here, the caracal is a generalist and apparently quite flexible, or plastic, in the way it hunts, enabling it to exploit a wide variety of prey types. On the other hand, Iberian lynx appear to be specialists with relatively inflexible prey-capturing techniques. This smallish lynx mostly lives on European rabbits, which constitute from 80 to 100 percent of its diet, and does not survive where introduced diseases have extirpated the rabbits. It does not, for example, switch to killing hares, which, even though they live in the same general areas, tend to live in more open fields, while rabbits live at the edges between shrub and fields. Occasionally these lynx do kill a deer fawn, and they catch and kill red-legged partridges where these are available in mountain habitats, but it remains to be seen whether they can live on red-legged partridges alone (see *What Are the Most Endangered Cats?*).

DO CATS HUNT INDISCRIMINATELY?

Biologist Hans Kruuk (1986) reviewed the studies of felid food habits and found that the 12 cat species in his sample hunted an average of four different prey types, compared with canid species, which hunt six or more. Kruuk arranged prey into the following categories: vegetable (including fruits and leaves), carrion, invertebrates, fish, amphibians and reptiles, birds, small rodents, lagomorphs (rabbits and hares), larger mammals weighing less than 50 kilograms, and mammals weighing more than 50 kilograms. The jungle cat (*Felis chaus*) was the most limited, eating only invertebrates and rodents. Lions and tigers ate larger mammals of both size classes and carrion. Cheetahs ate lagomorphs and larger mammals of both size classes. Pumas ate lagomorphs, larger mammals of both size classes, and carrion. Servals ate amphibians and reptiles, birds, and small rodents. Caracals ate birds, rodents, lagomorphs, and larger mammals weighing less than 50 kilograms. Bobcats ate rodents, lagomorphs, larger mammals less than 50 kilograms in size, and carrion. The African wildcat subspecies (*Felis silvestris lybica*) ate invertebrates, birds, small rodents, and carrion; the European wildcat (*F. s. silvestris*) added lagomorphs to this list. Taken together, Canada and Eurasian lynxes ate birds, rodents, lagomorphs, larger mammals of both size classes, and carrion. Leopards ate everything but vegetables, invertebrates, and fish. None of the cats in this sample ate fish, but this prey type is known to be eaten by flat-headed cats and fishing cats. None in the sample ate vegetables either (see *Why Don't Cats Eat Fruits and Vegetables?*).

The menu of big cats is largely composed of various kinds of hoofed mammals; in general they prefer prey about the same size to about twice as large as they are. John and Charles McDougal demonstrated that tigers in Nepal selected the largest ungulate available—sambar deer—even when they were not the most abundant prey available (Seidensticker and McDougal 1993). Ullas Karanth and Mel Sunquist found that tigers in south India also selected the largest prey available, including a very large wild cattle species called gaur, or Indian bison, in which males may weigh in excess of 1,000 kilograms (Karanth and Sunquist 2000). However, in both studies, tigers also took smaller prey, so the prey size averaged about 100 kilograms. Leopards in this same study site killed prey that averaged about 40 kilograms, with the largest weighing 175 kilograms.

Tigers appear to search a habitat in a way that will put them in position to kill a gaur or a sambar, but if they meet some other prey animal first, they will kill it if they can. In the Karanth and Sunquist study, tigers preferred gaur but avoided langur monkeys. Langurs live in groups and are constantly on the lookout for large cats, sounding their loud alarm call when they spot one and thus alerting other prey to the danger. Leopards seem to actively avoid wild pigs.

Cats in the medium-size range are all over the butcher shop, so to speak. Pumas living in western North America prey mostly on mule deer, white-tailed deer, and elk, which can be seven times the body weight of the cat, but when these big packages of protein are scarce, pumas settle for small rabbits or hares and ground squirrels. In the South American rain forest, where large prey are always rare, pumas survive on small rodents, bats, opossums, and lizards. In the cool hills of Chile, in western South America, they kill guanacos (smaller than but related to camels) and introduced European hares. Snow leopards eat 40- to 60-kilogram blue sheep (also known as bharal) as well as marmots (large rodents) and pheasants.

We have records of what leopards eat in many habitats from across their wide geographical range. Where John studied leopards in a Javan rain forest, they ate mostly primates (Seidensticker 1983). Very few ungulates of any size live in this habitat. In African and Asian habitats, where there are ungulates in the 20- to 50-kilogram range, they are the focus of leopards even if primates are reasonably abundant. Leopards seem to spend little time hunting smaller prey, although if something presents itself, these cats will kill and eat it, so small prey always appear as kind of a background to the leopard's diet. In the southern African Kalahari Desert, leopards take ungulates and also kill genets, springhares, bat-eared foxes, black-backed jackals, aardwolves, and porcupines. Basically, in this harsh environment, leopards go after whatever they can find. Leopards also specialize in killing jackals and dogs when there are available, as many dog owners in leopard range know. They will kill dog after dog, jackal after jackal, and not much else. This is a manifestation of leopards' ability to switch to and focus on the most readily available prey.

At the other extreme, the small prey of small cats, those weighing less than about 18 kilograms, consist mostly of diverse species of small rodents (mice and rats), rabbits or hares, and birds. All small cats studied thus far seem to include one or more species of rodent as a significant part of their diets. Lynx species have a special affinity for rabbits and hares. On average, snowshoe hares form about one-third of the diet of the Canada lynx, but in winter in Quebec, for instance, hares form as much as 85 percent of this cat's diet. Miscellaneous small mammals such as bats and hyraxes as well as snails, snakes, lizards, frogs, large insects, and spiders may round out a small cat's diet.

Any cat living near human settlements will make meals of domestic animals. Tigers take cows and water buffalo; snow leopards kill domestic sheep; leopards kill goats, calves, and dogs; small cats eat chickens and ducks. The bigger cats—tigers, lions, leopards, and, rarely, jaguars and pumas—also sometimes kill and eat people (see *Do Cats Attack and Kill People?*).

WHY DON'T CATS EAT FRUITS AND VEGETABLES?

Facultative and Strict Carnivores
Some species in the Carnivora live on a mixed diet of meat, fruits, seeds, and vegetables. These species are omnivores, or what nutritionists call facultative carnivores, and do not require meat outright. Other carnivores, including the cats, are called strict carnivores, or hypercarnivores, and cannot survive on a mixed diet; they must eat other animals. What's the difference? Animals' bodies fail to synthesize some essential molecules that are required by virtually all species, so these must be obtained from food. They include vitamins, essential amino acids, and some unsaturated fatty acids. Facultative and strict carnivores differ greatly in their nutrient requirements and the metabolic pathways through which their bodies obtain these essential molecules. Facultative carnivores have diverse metabolic pathways to digest their foods and can obtain their essential molecules from both plant and animal matter. Cats, as strict carnivores, have a narrow range of metabolic pathways by which to digest foods and obtain the essential molecules; thus, they must always eat other animals.

Proteins, Essential Amino Acids, and Vitamins

Animals must acquire energy, protein, water, minerals, and vitamins from their environment. In addition, they require other items such as fatty acids for metabolic function. An animal's protein requirement is linked to its nitrogen and essential amino acid requirement. Proteins are major constituents of an animal's body, so they must be in continuous supply for the animal to maintain its body and functions. All proteins are composed of about 20 to 25 different kinds of amino acids, some of which can be produced by the animal, some of which must be ingested. Those that must be ingested are called the essential amino acids; those that are produced by the animal are called nonessential amino acids. Mammals with simple digestive tracts, as carnivores have, usually require 10 essential amino acids. Mammals with complex digestion, as herbivores (rabbits, rodents, deer, antelope) have, require fewer essential amino acids because they can synthesize some entirely by themselves and some in conjunction with the symbiotic bacteria and protozoa that live in their digestive systems.

A mixed diet of fruits, vegetables, and meat provides omnivorous carnivores, such as dogs, with as much protein as they need, but cats have a much higher protein and nitrogen requirement for their maintenance. Adult cats require at least 19 percent protein in their diets compared with 4 percent required by adult dogs. Cats cannot live on lean meat alone for weeks on end, though. They must have fat components, which they get from eating the entire bodies of their prey. However, despite their high protein requirement, cats have a limited ability to regulate the enzymes that break it down, leaving them with a very high obligatory nitrogen loss. In essence, they leak nitrogen during protein digestion. They do not have the capacity to adjust their metabolism to conserve nitrogen, so they experience a high loss of it even when they feed on diets relatively low in protein. Dogs, on the other hand, can adjust their metabolism to conserve nitrogen when dietary levels are low.

Much of what we know about the digestive metabolism of dogs and cats comes from the problems that arise when owners feed their pets deficient diets. For example, a by-product of protein metabolism is ammonia. In order to convert ammonia to urea, cats require the essential amino acid arginine in their diet. A diet deficient in arginine leads immediately to ammonia toxicity in cats.

Cats also require high levels of the essential amino acid methionine, thought to be necessary for the production of taurine, and are unusual among mammals in that they need a dietary source of taurine. When domestic cats are fed synthetic diets devoid of taurine, their resulting low levels of it affect the functioning of their retinas: they lose the ability to resolve visual detail.

Another sight malady is related to a deficiency in vitamin A. Most mammals also produce vitamin A themselves, which is needed in turning beta-carotene into retinol, a step in the production of light-absorbing pigments in the retina. They eat plant foods to acquire the necessary beta-carotene. Cats, however, do not produce their own vitamin A, so they cannot use beta-carotene directly from plant sources; without a dietary source of both vitamin A and beta-carotene, cats suffer night blindness. Wild cats obtain both from animal fat and organ meat. However, a diet concentrated in vitamin A can be toxic. This vitamin is fat-soluble (meaning it is stored in the fat and used only as needed), so it can build up to dangerous levels in the body, which could happen in your cat if you were to feed it nothing but liver.

In contrast, the water-soluble vitamins flow out of the body constantly, so they must be replenished much more rapidly than fat-soluble vitamins. An example of a water-soluble vitamin is nicotinic acid, or niacin, in the vitamin B complex, which forms an essential part of tissue enzyme systems concerned in metabolism and the release of energy from cells. Cats cannot synthesize niacin in their body, and unless they obtain it through their animal prey, they suffer from reduced growth and ultimately die.

Carbohydrates

Cats appear to have no dietary requirement for carbohydrates, and they do not get significant amounts of carbohydrates in their animal prey diet, although these are an important source of energy for omnivorous carnivores. Cats lack the liver glycolytic enzyme that dogs and other mammals have to help digest carbohydrates, although cats do digest soluble carbohydrates by other pathways. They maintain what nutritional physiologists call high gluconeogenic enzyme activity, in which glucose is produced from protein. Cats, like all animals, require glucose as blood sugar, which provides a constant supply of energy for their bodies' cells.

Fats

Fats provide 2.25 times more energy than protein or carbohydrates do. So the animal fats in cats' prey are a very significant source of energy for them. Fats also provide the essential fatty acids called linoleic acid and arachidonic acid. Linoleate structures membranes for proper growth, fat transport, and normal skin and coat condition. Arachidonate is necessary for normal reproduction and blood platelet aggregation. Most other mammals, including dogs, can convert linoleate to arachidonate in their liver, but not cats. Their richest source of these essential fatty acids is the organ meat they eat.

DO CATS NEED WATER?

We have all watched a domestic cat squat over its water bowl and steadily lap up the water. The cat curls the tip of its tongue back, spoonlike, and flips the water back into its mouth, swallowing after every few licks. All the cats drink in this manner. How often they drink depends on local conditions and the cat's specific needs. Biologists have not compiled a list of the water requirements of the various cat species. We do know that water constitutes 99 percent of all molecules within an animal's body. Water makes up between 17 and 85 percent of a neonatal mammal's body weight, generally a higher proportion than in adults. Animals obtain "free water" from drinking in streams, puddles, lakes, and the like. They also obtain "preformed water" from food; when food is metabolized into energy, metabolic water is a product of the oxidation that occurs. A domestic cat's kidneys are more efficient (about 2.5 times so) than those of humans in ridding the body of waste products, and they can concentrate urine more than human kidneys can. Fed a normal dry cat food, cats will drink 1.5 to 2 milliliters of water for each gram of dry food. This proportion (2:10) is near the proportion of water in prey animals. Domestic cats need or drink little free water if they are fed other animal tissues.

We know little about the free water needs of wild cats, but biologists have observed the frequency of drinking and have noted the other sources of free and preformed water in some wild cats. Tigers may drink every day. Tigers also live in mangrove forests, which have little free water that is not highly saline, so in those environments, biologists suspect tigers obtain most of their water from their prey. Lions and leopards drink every day when free water is available, or they can go as long as 10 days without drinking. On average, both lions and leopards drink free water every 3 days or so. Lions and leopards living in the Kalahari Desert of southern Africa eat tsama melons, and lions lick the dew off other pride members. But lions, as well as sand cats and black-footed cats, living in environments essentially devoid of free water apparently get all the water they need from their prey. Black-footed cats, however, eat puzzling quantities of grass, which may be a source of preformed water.

DO ALL CATS AVOID WATER, AS DOMESTIC CATS DO? CAN ALL CATS SWIM?

An old proverb says, "Never was a cat drowned that could see the shore." A few wild cat species live in harsh, dry environments with little free water available, but most live close to water, and free water is an important part of their environment. Some seem fully at home in the water, even if they are not regularly observed using it. John once maintained a female puma as part of his research. While on a walk, the puma suddenly, and for the first time in her life, plunged into a wilderness river that was ice cold and in spring flood. The puma swam across it without pause to investigate something on the other side that had caught her eye. When she emerged, she simply shook her feet as domestic cats do when they get them wet in dewy grass. The point is that you cannot interpret a wild cat's likes, dislikes, or abilities regarding water until you have actually watched the cat for quite some time in its natural environment.

You can, however, align wild cats along a continuum from those living with little or no water in their environment to those living around lots of water. At the driest end is the sand cat, which can live in true desert without any free water. The black-footed cat, Chinese mountain cat, Pallas' cat, and Andean mountain cat live in very harsh, dry environments. Can they swim? Probably, but no one has reported them doing so. Lions, which live in habitats that range from dry to virtually waterless, are good swimmers and drink after every meal when there is water to be had.

At the other end of the spectrum are cats that live in, or even prefer, wetland and riparian (riverside) habitats. These include fishing cats, flat-headed cats, and perhaps Geoffroy's cats, known in some parts of their South American range as "fishing cats," but Geoffroy's cats also live in the relatively dry, windswept habitats of Patagonia. Tigers and fishing cats are completely at home in mangrove forests that nearly flood twice a day from the tides, but tigers also live in tropical dry forests, where water is in short supply but sufficient to provide the tiger with water every day. Tigers and ocelots rest in water to keep cool during the hottest part of the day. Tigers, fishing cats, and flat-headed cats are strong swimmers and even swim submerged, enabling them to grab prey, such as a floating duck, from underneath. Tigers have been found swimming far from shore, crossing tidal rivers more than 5 kilometers wide. Jungle cats are also strong swimmers, often found in wetland habitats, where they sometimes dive to catch fish in their mouth. The jaguarundi, another strong swimmer, is sometimes called an otter cat, and the leopard cat was given its particular species name, *bengalensis*, because the first one captured by Westerners was swimming in the Bay of Bengal. Swimming ability may have enabled leopard cats to colonize offshore islands that other small cats have not.

Some of the wild cats, such as flat-headed cats and fishing cats, are so closely associated with ponds and streamsides that we should think of them as living in a linear, one-dimensional landscape rather that in a two-dimensional area, because their home areas stretch along the linear extent of the stream or pond bank, not over an area whose boundaries form a polygon.

Some cats, such as African golden cats, oncillas, margays, flat-headed cats, bay cats, and marbled cats, live only in rain forests, but some rain forest species, including Asian golden cats and clouded leopards, also range into moist tropical

Like this ocelot cooling off in a pond, most cats are quite at home in water, and none are known to be unable to swim. Lions, however, seem to dislike water as much as domestic cats do, and desert-living species, such as sand cats and black-footed cats, never have access to enough water for us to judge whether they like it or not.

forest habitats. Jaguars, strongly associated with streams and watercourses, are strong swimmers capable of crossing large rivers, and they may rest in streams in the heat of the day. Ocelots, jaguars, and jaguarundis live in rain forests, but they also live in dry areas or seasonally dry areas by staying in riparian habitats and wet coastal areas.

The serval is a cat of savannas and wet grasslands. Servals hook fish out of the water and don't mind getting their feet wet while they hunt frogs, waterbirds, and crabs. The caracal lives in dry forest edges and scrubby, dry thorn forests, as the rusty-spotted cat does, but both seem always to include water holes within their home ranges. Some cats, such as pumas, are strong swimmers but seem merely to tolerate being in water, while lions, even though good swimmers, avoid water, as domestic cats do. Leopards that live in some arid environments seem at ease in water, play in it, and swim well for distances of 900 meters or more, but, unlike tigers, they do not inhabit mangrove areas.

WHERE DO CATS HUNT?

The broad habitat preferences of cats tell us only in the most general way where cats hunt. To know specifically where an individual cat will hunt, we have to focus on how a cat uses its habitat to meet its needs. Within a general habitat, a cat has to hunt where its prey is and where it can catch that prey—some in trees, some in water, some by chasing, some by ambushing. At any given moment, 10,000 kilograms of prey support about 90 kilograms of carnivore; that's a lot of meat. A habitat has to produce these potential prey animals, and they have to be distributed in

As their name suggests, fishing cats are among the most water-loving of cats, finding most of their food, which includes fish as well as small mammals, in or along stream sides and ponds.

the habitat in ways that allow a cat to hunt and capture them efficiently. Adequate cover from which to hunt undetected by prey is important; so is the distribution of trails, edges of microhabitats, springs, stream banks, and other features that attract prey, thus making it easier for a cat to find them.

A cat's hunting behavior forms a five-part series that includes searching for prey, detecting prey, orientating and approaching prey, capturing and killing prey, and, finally, eating it. Different sets of stimuli guide each of these behaviors except for searching, which cats do without the aid of any direct stimuli to guide them to the exact location of prey. Hunger drives the readiness to hunt, and a cat must be hungry enough to motivate its searching for prey even through many failed attempts to capture it. A cat begins its quest for food by going to a known area where it has successfully found food in the past; biologists call this hunting by speculation. When it is hungry, your pet cat goes to where its food bowl is usually kept; a tiger walks slowly down a forest road; a feral domestic cat strolls down an alley. Likewise, many cats set off at near dusk to check places where they have experienced success before, such as at the habitat edges between forest and field, along streams, or at moist spring seeps in otherwise dry forests, where small mammals are most abundant and deer and wild pigs come to drink. Many people imagine that cats lie draped over a tree limb, waiting to pounce on prey moving below on the ground. You see this image in cartoons all the time. In fact, most cats hunt on the ground, although some arboreal cats actively hunt (not just lie in wait) in trees for prey species that also live there. In general, cats hunt where they are most likely to find prey they can capture.

The availability of cover influences where all cats hunt. Lions and cheetahs live in open grasslands, but they need the cover of grasses and other vegetation in their hunts. Lion hunting success varies from 12 percent where there is little or no cover to 41 percent where there are dense, riverside thickets. What a small cat experiences as adequate cover within a habitat isn't the same for larger cat species or even for different individuals of the same species. A leopard cat, fishing cat, leopard, and tiger view a square kilometer of habitat in Nepal entirely differently, because each is seeking its own limited range of prey, and where they can find and catch prey differs, as do the risks each faces from other predators in various habitat conditions.

In one Indian dry tropical forest, the most abundant wild ungulates were cow-sized antelope, called nilgai or blue bull, and sambar deer. Nilgai were slightly more abundant than sambar, yet a female tiger killed twice as many sambar as nilgai, because nilgai stayed in more open areas, away from trees and shrubs, often in groups, making it hard for a tiger to sneak up on them. (The Indian cheetahs that once lived there could have succeeded, however, because cheetahs begin their very fast pursuit of prey from farther away.) However, water sources were few in this dry region, and the tiger knew that sambar came to the few water holes to drink each night. Brush and other cover surrounded the water holes, making it easier for the tiger to ambush the sambar. Domestic cattle were several times more abundant than wild prey in the same area, and an adult male tiger always killed cattle, except during the monsoon rains, when lush grass allowed herders to graze their cattle close to their villages.

The number and distribution of good hunting spots—*sweet spots* to the cat—determine the size of a cat's home range, or territory, and thus the densities at which cats live. When a cat is a newcomer to an area, it spends a considerable amount of time exploring. Exploration decreases with experience as the cat succeeds in making kills. Biologists speculate that each cat has an internal mental map—a sense of what it is looking for, which it is continually comparing with the stimuli (good, bad, neutral) it receives as it explores, enabling it to learn easily and quickly where to hunt. While this mental-map hypothesis seems intuitively plausible, we have yet to confirm or reject it through actual experimentation. We do know that wild cats learn quickly. In Idaho's Salmon River Mountains, John found that young female pumas seemed to hunt all over the place when they first appeared in a winter home range, but the next winter, they were killing deer and elk in the same places they had killed them the winter before, indicating they had learned to hunt the sweet spots in this area (Seidensticker et al. 1973). This range knowledge is a tremendous benefit in improving their reproductive success, because a female puma (or any cat) can go directly to sweet hunting spots, without wasting time, and expect to make a kill, thus improving the chances of her litter's survival. This explains why many resident female cats are reluctant to change or expand their home areas, even when conditions change or adjacent territories become available because of the death of another resident female. An observer may know there is more prey in the adjacent area, but the female cat doesn't, and she doesn't have experience in that area that would allow her to take advantage of it. Female lynx, for example, use about the same home areas when hares are abundant as they do when hare numbers crash. They know the lay of the land and have the advantage of knowing the best hunting spots even in hard times.

Tigers inhabit tropical rain forests, mangroves, tall grasslands, mixed deciduous forests, reed jungles, riparian thickets, mountain forests, dry tropical forests, and temperate and boreal forests. What is critical to a big, mobile tiger's survival in these various habitats? The daily alteration and loss of habitats across the tiger's range motivated conservation biologist Dale Miquelle and his colleagues to find out what matters most to tigers, as a basis for future conservation action (Miquelle et al. 1999). They found that habitat selection by tigers occurs at several levels. First, tigers live in a particular geographical range, which once included large parts of Asia. Second and third, the home range of any individual tiger incorporates both a variety of vegetation types and particular vegetation types. Finally, the home range includes the kinds of individual prey animals that a tiger actually kills.

In the Russian Far East, Miquelle and his associates found that vegetation types per se did not predict the tiger's distribution, but they did predict the presence of key prey species. Most important was the distribution of those key prey species; they were the greatest predictor of the tiger's distribution, which was more closely associated with the distribution of red deer than with that of any other prey species, even wild pigs, which are thought to be a key prey species. However, combining the distributions of wild pigs and red deer increased the strength of the relationship with tiger distribution beyond that of either species alone. The message is: follow the prey, and the tiger will be there. Across the tiger's range, the density of this species is directly related to the density of its prey.

Depending on the cat's view of the world, the sweet spots in a cat's home range may include a spring, a small marshy area in an otherwise dry habitat, shrubs along the bank of a stream, a trail in a forest, the edge of a forest and grassy meadow, the base of a steep bank, or any number of other topographic and vegetative combinations that attract prey to where they are vulnerable to the cat. Prey concentrations constitute sweet spots in their own right if they are in places where a cat can stalk and kill without exposing itself to predation. For example, it doesn't do a snow leopard or a puma much good to ponder thousands of antelope or elk in broad, open valleys below them. The broad, open valley is wolf country, and wolves will kill pumas and probably snow leopards. To hunt successfully and avoid predators, a puma or snow leopard needs to seek the antelope or elk that frequent the broken terrain at the edges of the valley. This is the microhabitat that matters to these cats.

The makeup of the places where cats can kill differs with the size of the cat. For example, as we've seen, in the Russian Far East, red deer are a dietary mainstay for the tiger. In the Rocky Mountains of western North America, elk are the puma's mainstay. Elk and red deer are subspecies of the same species, *Cervus elaphus*, and are about the same size, but a tiger is about four times larger than a puma. The environmental circumstances in which a female puma can approach and kill an animal seven times her own size are more exacting than those required by the larger tiger to kill prey of that same size.

Sweet spots that attract prey and increase a cat's hunting success can also become death traps for the cat. Besides the aforementioned open valleys, roadsides are a good example. Plants that attract potential prey species such as deer and rabbits grow along roads; the prey attract cats. However, they also place the cat in danger of being killed by moving vehicles and by poachers, who know they can find and kill cats visiting these areas at night.

WHEN DO CATS HUNT?

Most felids can and do hunt both day and night. Cats, thought of as primarily nocturnal, readily hunt during the day if prey become available then. What seems to drive their activity is the need to hunt when their prey are most vulnerable. It does not make sense for a cat to hunt always at the same time of day, because their prey would soon learn to adjust their own activities accordingly. "Predators are known to synchronize their predatory activity with the main activity of their prey. This in turn opens up the possibility for the prey to develop defenses based upon a daily rhythmicity," wrote ethologist Eberhard Curio (1976, 34). But cats can also take advantage of rhythmicity in prey activity that is determined by other predators.

Ocelots find their prey by stealthy walking both day and night, with most of their activity in the late afternoon and night. Conservation biologist Louise Emmons and her colleagues published an enchanting paper titled "Ocelot Behavior in Moonlight" (1989), based on their observations of this predator's hunting in perhaps the most untouched tropical rain forest in South America. These biologists found that the number of ocelot tracks on trails and on stream- and pond-side beaches decreased as the moon waxed, but radio-tracking showed that the time ocelots spent active, the total distance they traveled, and their activity patterns did not differ significantly on moonlit nights versus dark nights. In this pristine rain forest, trails were narrow, well-worn, and followed natural boundaries such as gullies and river edges. In general, ocelots used trails, probably because they could walk quietly and easily on them and could see prey from farther away, making it easier for them to surprise that prey. They also could keep drier when the vegetation was wet. But on moonlit nights and during the day, the ocelots avoided these open areas, hunting instead in denser cover. Brighter conditions may reduce an ocelot's hunting success by making it harder to approach prey undetected. Likewise, spiny rats, a primary prey species, were equally active on dark and moonlit nights but less visible on trails on moonlit nights.

These biologists found that the movements of ocelots were unpredictable from one day or night to the next because of the unpredictability of capturing prey. An individual might walk for 12 straight hours, then spend more than a day in the same place. Over time, however, it became clear that ocelots were most active just before and during the first hours of night. Tropical small mammals are nearly always nocturnal, which influences the activity of their predators. But prey must adjust their activity in relation to more than a single predator species. Emmons and her group believe that diurnal and nocturnal birds of prey, or raptors, determine the activity patterns of small mammals: "Birds of prey seem most likely to exert the greatest selection against the activity of small mammals in moonlight, as increasing light may allow them to hunt in places where they cannot hunt in the dark" (Emmons et al. 1989, 239). Ocelots adjust their own hunting times to take advantage of the activity rhythm of small mammals, as set by the raptors. But they also have to adjust their activity to other predators that kill ocelots—pumas, jaguars, and anacondas, for example—which perhaps keeps ocelots off moonlit trails, where they may be more easily seen. Jaguars, in contrast, which are the largest carnivore in their range, walked on beaches and trails in their study area at any time.

Using radio telemetry, scientists have documented the activity of several other wild cat species. In a long-term study, lions spent about 80 percent of their time inactive—sleeping, lying around, doing very little. They were active mostly at night, when they spent 10 to 15 percent of their time traveling and only 5 to 8 percent of their time hunting and eating what they killed. Lions have been observed to make 88 percent of their kills at night in the Serengeti. In a Nepal study, tigers were active and moved primarily at night, although some daytime activity and movement were not uncommon. Cheetahs were active in the morning and afternoon, when predators and competitors such as lions were inactive, resting during the heat of midday. Pumas in the Rocky Mountains were more active during the day in summer, when they preyed on diurnal Columbia ground squirrels, which hibernate underground in the fall and winter. Pumas killed deer and elk in the evening, at night, and in midmorning. Leopards living in South Africa's Kruger National Park were active about 78 percent of the time at night and 58 percent of the day. Clouded leopards in a Thailand rain forest were arrhythmic but more active at night and at

Cats hunt where they know they can find food and alter their hunting behavior with changing circumstances. Ocelots, for instance, hunt on dark nights along easily traveled trails on which they can spot prey from a distance, but on moonlit nights or during the day, when prey might see them first, they hunt instead in thick cover.

dawn and dusk (which is called crepuscular behavior). Ocelots living in Venezuela's llanos were active 12 to 14 hours a day. They were significantly more active at night than during the day, but diurnal activity increased in the wet season. Servals living in Tanzania's Ngorongoro Crater were mostly crepuscular, with a main rest period during the hottest time of the day. Fishing cats photographed by camera traps were active both day and night in Colombo, Sri Lanka (see *How Do Scientists Study Cats Living in the Wild?*).

Seventy percent of the movements of male Eurasian lynx living in Poland's Białowieza Primeval Forest occurred at night, but females were active as long during the daylight as during the night. Lynx activity was determined by their hunting success: on days when they hunted but failed to kill, they were active for 12 hours or more; when they made a kill, activity was a minimal 1.6 hours a day. Leopard cats living on the island of Borneo were primarily nocturnal with some sporadic but very limited daytime activity. Jungle cats have often been seen hunting during the morning and late afternoon. Geoffroy's cats living in southern Patagonia were primarily nocturnal with two activity periods each night. Margays ranged from strictly nocturnal in some areas to equally active in day and night in others. Sand cats avoided the heat of the day, and hunting began just before dusk and continued through the night. Kodkods living in southern Chile were active 85 percent of the time and were equally active day and night. Black-footed cats hunted only after dark in the South African coastal shrub.

HOW FAR DO CATS TRAVEL IN A DAY OR NIGHT AND HOW FAST?

How far a cat can travel in a day or night depends on its size. Because cats are secretive and move in cover, frequently under the veil of darkness, it is normally impossible to follow them. Radio telemetry helps, but even if you sample a cat's activity and movement every 15 minutes, you can't be sure how far it has actually moved—you know only the distance between the two sampling points.

Conservation biologist Paul Beier and his associates conducted a detailed study of puma activity and movements in the Santa Ana Mountains of southern California (Beier et al. 1995). Through radio tracking and other observations, they found that pumas are overwhelmingly crepuscular and nocturnal. Those observed stalked or sat in ambush for periods averaging 0.7 hours and then moved an average distance of 1.4 kilometers over the next 1.2 hours to get to another site. This pattern was repeated six times a night for a total travel distance of about 10 kilometers per night. When a puma killed a small mammal, it stopped for 4 to 6 hours; when it killed a large mammal it stopped for 2 to 5 days. Average travel speed was 0.8 kilometers per hour but reached 1.5 kilometers per hour during the longest travel bouts. (The normal walking speed of a captive puma was 1.6 kilometers per hour.) The maximum speed these biologists observed was 4.6 kilometers per hour, well below 8 to 9.6 kilometers per hour, which is the transition between a walking and a running gait. On average, the pumas they tracked traveled for 25 percent of each 24-hour period.

We have some information on the movement rates of other cats based on intensive observational studies aided by radio telemetry. In a South African study, black-footed cats were observed to have three hunting styles. They traveled at 2 to 3 kilometers per hour when they "fast hunted," which flushed prey as they went. In slow hunting, they traveled at 0.5 to 0.8 kilometers per hour, moving through the grass or bush in a stalking, sinuous way. In ambush hunting, they just sat and waited for prey to appear. Lions traveled on average between 5 and 9 kilometers per day in the Serengeti and Ngorongoro Crater, depending on the season and habitat; but in 12 hours one female traveled 20 kilometers between a den where her cub was hidden and the location of the rest of her pride. Leopards walking at a speed of 3 to 6 kilometers per hour easily covered 12 kilometers in a night; where prey density was low, they walked 16 to 17 kilometers per night and up to 33 kilometers in a 24-hour period. Chitwan tigers were on the move 10 to 12 hours a day, mostly traveling at night for about 7 to 10 kilometers, roughly at 0.7 kilometers per hour. One adult male traveled as far as 30 kilometers a day. Tigers in India's Kanha National Park traveled 16 to 32 kilometers a night; in the Russian Far East, tigers traveled 15 to 20 kilometers in a 24-hour period. Iberian lynx traveled 8 kilometers each night. Bobcats traveled between 4.8 and 11.2 kilometers per night. Canada lynx nearly doubled their average nightly travel distance, from 4.8 to 8.8 kilometers, in response to a decline in snowshoe hares. Ocelots moved at a slow and steady pace of close to 0.3 kilometers per hour when hunting and 0.8 to 1.4 kilometers per hour when traveling from a resting site to a chicken kill. Borneo leopard cats were observed to move at an average rate of 0.1 kilometers per hour, and their average one-day movements were 0.69 kilometers.

When they want to, these little cats can move. Just at dusk John watched a leopard cat trotting at a speed in excess of 5 kilometers per hour along the edge of the beach at high tide in a national park in Bali, Indonesia. The high tide had narrowed the strip where potential prey were concentrated on the beach, and the leopard cat seemed to be using this narrow window of opportunity to cover as much of the beach edge as it could in as little time as possible; it could see a long way ahead and was fast hunting.

John radio-tracked a young female bobcat during one spring as she moved about her home range in rural Virginia. After a month or so, he learned that she predictably moved throughout her entire area of 10 square kilometers every 4 days. Emmons reported that the nightly paths of adult ocelots seemed deliberately chosen. They walked in smooth lines or long loops, rarely backtracking and covering a major part of their territorial boundaries each night. They chose different pathways on sequential nights, so the entire home range boundaries were visited every 2 to 4 days (Emmons 1988).

After a puma left a kill site in the Salmon River Mountains, John found that it almost inevitably traveled rapidly to the farthest extent of its home area and then slowly worked its way back through the home area until it killed again. A female tiger that John radio-tracked spent her days lying in very thick grass and near dusk walked directly to specific hunting areas. Once arrived at a hunting area, she slowed down and deliberately worked her way through it. Her home area was only 20 square kilometers, and she could cross it in less than 2 hours (Seidensticker 1976).

HOW DO CATS FIND AND PURSUE PREY?

Searching by Expectation

John watched the hunting sequence of a tigress, the first ever radio-collared, in the tall grasslands and riverine forest in Nepal (Seidensticker 1976). About a half hour before dark, her radio signal warned John that, very directly and rapidly, she was approaching the tree blind he was sitting in. She had been walking down a dirt road, but as she came into view she turned off the road and abruptly slowed down to follow a narrow trail that snaked through the 4-meter-tall grass. She walked deliberately, her head held high. After about 15 meters, sensing something off to her right, she halted and turned toward it. She could not see what it was behind the tall grass, but she may have been able to tell what it was by its smell and sounds. From his hide, John then saw a hog deer abruptly jump a few times away from the tiger, which was about 10 meters away. She never saw the hog deer but clearly knew it was there and heard it jump away. Making no attempt to pursue it, she immediately turned back and, head held high, proceeded down the trail.

As she neared a large clump of tall grass that obscured the trail directly below John's hide, she paused, raised her head higher, sniffed the air, stared, and then darted around the tall grass. As she went around one side, a wild pig came out the other side of the clump, about 5 meters in front of her. John had not seen or heard the pig, and the tigress could not have seen it either, but she had sensed the prey (heard or smelled it) and gone after it. However, the pig had also sensed her, so, again, she did not give chase and just walked off into the forest, disappearing. Within 5 minutes of her arrival at this hunting area, having come from a lay-up site where she had spent the day with her cubs, she had stopped and orientated toward one potential prey and pursued another without seeing either of them but hadn't tried to capture them. She did, however, make a kill later that night.

It was striking how completely aware of the circumstances she seemed to be and how she immediately assessed when to abandon the hunt after the deer and the pig prey had detected her, even though they remained nearby. The advantage she seemed to seek was that of surprise. The area she was hunting in contained more than 50 potential prey animals per square kilometer, so she could expect plenty of other opportunities in the night to come.

Cats primarily hunt by sight, but these observations show us that hearing and smell also play an important role in finding prey. How cats find prey has puzzled biologists, because the dynamics are so difficult to observe without disturbing them, and the situation is also fraught with ambiguity. It is easy to determine what a cat has killed and to compare this with what prey are available. But this doesn't tell us how cats search for and find prey. We know that cats go to and walk along places where they expect to find prey. They are always scanning and moving to enhance their perceptual field, for example, by walking along roads and trails where they can see animals from farther away than is possible in thick vegetation. In our experience, cats know what they are looking for—and they may multitask. When snow-tracking pumas, John found that one walked around hillsides in a way that took it both to places where elk might be feeding out of the wind and to places where it might meet a mule deer eating mountain mahogany at the edge of some bluffs (Seidensticker et al. 1973). While traveling up a creek bottom between such sites, a puma would suddenly veer off the trail to pass slowly under a rock overhang (a good place for a pack rat) or swerve cautiously to a stream side (a good place for a beaver). Pumas searched for prey animals in places their experience told them prey were likely to be. They moved in a way that seemed to maximize the chance of finding a mule deer or an elk but did not ignore opportunities to find other prey as they traveled along.

It is interesting to compare cats with weasels, fishers, and other mustelids. Cats search for active prey, prey that is feeding or otherwise distracted. Cats are attuned to detect the slightest movements that might alert them to prey. Mustelids, in contrast, search for inactive prey. Otters don't chase swimming fish, they concentrate on resting fish and slow-moving bottom dwellers, and they catch rabbits inside their burrows. Fishers don't lie in wait, nor do they usually chase snowshoe hares, their most important food; instead, they go where hares are likely to be found and forage for them. According to Malcolm Coulter, "In hare country, they dart beneath every low-hanging bough, fallen tree top, log or similar place where a hare might be sitting . . ." (cited in Powell 1993, 118). Weasels and martins, too, forage in this manner and also forage under the snow for mice and voles in their nests and burrows. This probably is the more primitive hunting mode among the Carnivora, and the stalk and rush of felids is more advanced.

The Stalk

Once a cat detects a prey animal, it must first decide to approach and then approach successfully. Cats excel in the approach, and stalking is characteristic of cats—it is one of the primary things that make a cat a cat. Biologists are always impressed by how rarely you see wild cats when you walk in the woods. Sure, wild cats live at very low densities compared with their prey, and encounters are unlikely on chance alone. But there is more to it than that. A cat nearly always senses you first and withdraws or simply stops and blends in with the background. And so it is when a cat is hunting. A wild cat's life depends on its ability to sense a prey animal before the prey senses it and then approach undetected to a distance where it can successfully rush in, seize, and kill. The sensory center for predatory prowess resides in the superior colliculus, a part of the midbrain, which acts as a multisensory relay system to receive visual and tactile information and relay it to the spine via the tectospinal neural tract; nerve cells in this tract are larger and more numerous in cats than in their prey.

Cats have an amazing ability to accelerate—explode—in their final movements to overtake prey. But they have no physiological ability for sustained running, as dogs and their prey do. Even cheetahs are limited in how far they can purse prey, because they cannot get rid of the body heat that rapidly builds up while running. Even a moderate rise in body temperature is deadly. So a cat has to close in and kill in very short order, or it has to start the hunting sequence all over again.

The final rush involves a matching of the sprinting abilities of the predator and the prey. Among cat species, differences occur in the relative length of the stalk and in the final rush. The smaller felids often end their stalk with a pounce, descending vertically with the forefeet onto small prey. Or they rear up to snatch a flushed bird out of the air. In the Serengeti, leopards, lions, and cheetahs hunt Thomson's gazelles; a cheetah will run at one from an average distance of 300 meters, a lion from 30 meters, and a leopard from 10 meters. Unusual for cats, cheetahs do not always stalk before making the final rush; leopards make very elaborate use of cover before their final rush.

European wildcats and feral domestic cats use two different strategies for catching rabbits and rodents. In the mobile strategy, the cat roams until it sees and can stalk a rabbit; in the stationary or ambushing strategy, the cat crouches next to a warren, waiting for a rabbit to emerge. Larger and more aggressive cats more often use the mobile strategy and have larger home areas, whereas cats with smaller home areas use the stationary strategy.

At the Smithsonian's National Zoological Park, we have watched how large cats respond to potential prey stimuli, usually in the form of groups of visitors. The larger cats nearly always focus their attention on the smaller kids in a group of adults and children, especially the ones that are fidgeting and bouncing around. The cats are responding to the contrast, and this is in part how they select their prey. Given a choice, predators in general select prey that differs from the rest of the group in looks, for example, a smaller animal; or in movement, for example, the one limping; or in spatial relationship, for example, one that is just a little apart from the rest of the group, closest to the trees or bringing up the rear. We also have learned that a large cat's response to people depends on how they present themselves. A person standing tall does not attract the same attention as a person kneeling, squatting, or on his or her hands and knees.

This jaguar is performing the most characteristic of cat behaviors: the stalk. Quietly and with seemingly infinite patience, a cat that has detected prey crouches low to the ground and slinks toward it, using every available bit of cover and never blinking an eye. It freezes if the prey looks up and advances again when the prey looks away. When near enough to rush it, the cat may pause, waiting for the best moment to attack.

Unlike chasing predators, such as wolves or wild dogs, stalking predators do not "test" a herd by putting it into flight and then selecting the individual that falls behind or veers off to the side. This is what is popularly known as weeding out the weak and the sick. The manner in which cats hunt does not lend itself to such testing. No difference was evident in a comparison of the physical condition of two samples of mule deer, one set killed by pumas and the other selected at random.

Remaining inconspicuous is all-important to a cat, and when detected, cats abandon the hunt. But none of the cat species seem to try to avoid letting their prey smell them by considering wind direction when they stalk. Once a stalk begins, the cat crouches low to the ground. When the target prey turns away or looks down to feed, the cat advances rapidly, then freezes, advancing again when possible, until it judges itself close enough to make the final rush, which is determined on the basis of its previous experiences.

The Final Rush

For most cats, the rush part of the hunting sequence is guided by direct visual cues. However, we know that some species, including servals and sand cats, seem to be guided by auditory cues alone during this stage. Some smaller cats, such as servals, lunge at prey, launching themselves like missiles, to come down from above, pin the prey to the ground with the forepaws, and then bite to kill it. Aadje Geertsema described this:

> When the serval detects a prey animal in the surrounding vegetation, the ears pick up and pinpoint the exact location, while the body remains tense and still. This can be accompanied by "treading" of the rear feet and "tail-twitching." At this stage the serval moves forward very cautiously for a few meters, often with one foot poised in the air, depending on the distance of the prey. He then stops again or immediately performs a single pounce or a series of pounces. A single pounce may span 1 to 4 meters and may be over a meter high. . . . If a pounce fails, servals may switch over to a techniques of bouncy, zig-zag pounces while following a rodent trying to escape. (Geertsema 1985, 579)

She watched nearly 2,000 pounces, of which 49 percent were successful.

The stalk is followed by the rush—an explosive burst of speed that closes the distance between predator and prey. To be successful, jaguars and other cats must outaccelerate their prey during the rush; therefore, their ability to determine how closely they should stalk before rushing is critical.

John Elliot and his associates conducted a comprehensive analysis of a lion's behavior during the final rush in the Ngorongoro Crater, the same place where Geertsema watched servals pounce (Elliot et al. 1977). According to Elliot, a lion's hunting success depends on the intended prey's not seeing the predator's approach and also on the lion's ability to judge when it is close enough to rush. This involves matching its sprinting ability to that of the prey. The lion must launch its rush and then accelerate faster than its prey to close the gap. If lion and prey start at the same time, the lion will be outaccelerated by a wildebeest within about 8 seconds, by a zebra within about 6 seconds, and by a Thomson's gazelle within about 4 seconds. This means that a lion's success drops off the greater the distance between predator and prey when the attack is launched. The lion has no chance of capturing a gazelle if when the rush starts the distance between them is greater than 13 meters; it has less than a 20 percent chance of capture at distances greater than 30 meters from zebras or wildebeests. Lions can, however, increase their hunting efficiency by pursuing very young and enfeebled animals, because they have greater stamina than such prey.

HOW FAST CAN CATS RUN?

Cheetahs epitomize the running cat. Although they are well known as one of the fastest land mammals, just how fast cheetahs run is not known with certainty, because speed is difficult to measure under field conditions. Estimates range from 90 to 104 kilometers per hour, but the exact number is not important; what is important is that it's fast enough for cheetahs to capture speedy gazelles. In turns out, though, that the key to cheetahs' success is not speed per se but acceleration. From a complete standstill, cheetahs reach speeds of 75 kilometers per hour in just 2 seconds, so fast that human eyes see a blur when a cheetah takes off. This is what gives cheetahs, and most felids, an edge over their prey.

Lions can reach speeds of 48 to 59 kilometers per hour but, according to George Schaller (1972), have little stamina and pant after running fast for 100 meters. At the other size extreme, small sand cats reportedly sprint at more than 30 kilometers per hour. Jungle cats reach 32 kilometers per hour, and domestic cats 48 kilometers per hour.

HOW FAR CAN CATS JUMP?

R. McNeill Alexander wrote, "Running is a series of leaps" (1992, 25). When an animal is running, there are stages when all feet are off the ground, when it is, however briefly, essentially sailing through the air. Leaping, or jumping, is the most

Like all cats, snow leopards can run fast but only for a short distance. Cats are sprinters, not marathoners, and wear out quickly. As a result, cats do not run down prey; if a cat's rush fails to capture its intended prey, the cat gives up and waits for another chance to kill.

typical movement of cats; propelled by relatively long, powerful hind legs and strong thigh muscles, some cats can leap lengths that Olympians might envy. Keep these numbers in mind: the Olympic record for a long jump is 8.9 meters; for a high jump it is 2.39 meters. Basketball great Wilt Chamberlain could jump about as high as he is tall, just a bit over 2 meters (7 feet, 1 inch).

Other things being equal, how high an animal can leap is proportional to its body mass. But clearly other things are not equal among the cats. Usually lighter in mass than an adult human male, a puma's record long jump is 12 meters, its high jump an astonishing 6 meters. The puma's hind legs are very long for its size, and its jumping ability is undoubtedly an advantage in the often steep terrain it occupies. Tigers, with greater mass but shorter hind limbs than pumas, are reported to jump 8 to 10 meters; leopards leap 6.6 meters or more; and snow leopards, at least 6 meters.

Among the small cats, the acrobatic and arboreal margay can jump 2.5 meters vertically and nearly 4 meters horizontally. Servals leap vertically to pluck birds

Cats are champion long jumpers. Propelled by strong hind legs, a Canada lynx can jump—essentially sail through air—more than 7 meters to pounce on a hare.

midair; caracals do this also and have been reported to jump 3.4 meters straight up a wall. Similarly, a tiny black-footed cat can jump 1.4 meters into the air and 2 meters horizontally. Servals, which often pounce on their prey, perform what is called a capriole jump, similar to the pounce of a fox. Writing about the red fox, Alexander describes it this way, "At take off, the muscles of the hind legs throw the animal into the air: most of the work is done by the big muscles of the thighs" (1992, 1). In servals, a single pounce may cover 3.6 meters horizontally and 2 to 3 meters vertically.

ARE CATS PERFECT HUNTERS?

Cats have to work for their meals and often miss them. By closely following habituated black-footed cats, biologist Alexander Sliwa (1994) found they had about a 60 percent success rate. By reconstructing hunting sequences through snow tracking, wildlife ecologist Maurice Hornocker (1970) judged that 82 percent of puma hunts were successful, but he did not count hunts when the puma quit. In one study, Serengeti lions hunted 145 times and killed 34 times (a success rate of 25 percent) over 12 4-day observation periods (Hanby et al. 1995). P. Stander and his

Cats are often described as perfect hunters, but they miss kills surprisingly often. In a Serengeti study, female cheetahs killed hares almost every time they tried to but were successful in only 45 percent of attempts to kill Thomson's gazelles.

associates (1997) found that Namibian leopards averaged 1 kill for every 2.7 hunts, with an overall success rate of 38 percent for all prey species combined. Andrew Kitchner (1991) reviewed published information on hunting success and found enormous variability. Tigers in Chitwan succeeded in killing, on average, in only 1 or 2 hunts out of every 10 (Sunquist 1981).

Servals studied in Tanzania were successful in about half of all pounces, although this average may conceal much variation among the prey species: 54 percent of servals' pounces on insects met with success, 49 percent of pounces on rodents, but only 23 percent of pounces on harlequin quails (Geertsema 1985). Canada lynx hunting snowshoe hares on firm snow killed in 24 percent of hunts but in only 9 percent on soft snow. Success among lions varied with the amount of cover—41 percent with dense cover versus 12 percent with light or no cover—and was higher at night than during the day and higher on moonless nights than on moonlit ones. Tim Caro's (1994) detailed analysis of hunting success among Serengeti cheetahs reveals additional sources of variation. Looking at females and young, he found that adults improved at catching hares and baby gazelles as they matured, eventually reaching a success rate of almost 100 percent. When hunting other, larger prey,

adults were more successful than young adults, and adolescents almost never succeeded in killing larger prey. Adolescents and young adults tended to be seen by
prey far more often than adults were, although this was the leading cause of failure
for all three age groups, as it was for adult males alone. A cat's reproductive circumstances also affect its hunting success. Lactating females (those with cubs less
than about 4 months of age) were more successful than females without cubs; cubs
traveling with their mother often accounted for failed hunting attempts, because
their inappropriate behavior alerted the prey to danger.

If cats have evolved to find prey, then prey has evolved to foil predators. Behav-
iorist Valerius Geist (1998) points out that the evolution of deer, for example, is
written in the diversity of their various antipredatory strategies. Prey countermea-
sures take many forms: they bunch, face, mob, bound, or run away. John once
watched a leopard rush a group of hog deer and disappear into their midst; one hog
deer jumped above the others, like a cork out of champagne bottle, and the rest of
the deer ran in all directions at once. The leopard, alone and without a kill, paused,
then trotted off.

HOW DO CATS KILL PREY?

Primitive and Advanced Carnivores
The most basic way for one animal to kill another is simply to bite the prey, letting
the tooth puncture kill it outright. Along with this killing bite, the predator may
also repeatedly toss and throw the prey to damage its spinal column. For this to
work, the predator usually has to be considerably larger than its prey. Grasping prey
in the forepaws, which is a relatively modern evolutionary advancement, is associated with delivering a precisely aimed killing bite to a particular spot on the prey's
body, usually the prey's skull and/or neck. Holding the prey in the mouth while
shaking it vigorously from side to side is another advance, derived in part from the
primitive tossing action.

How a cat seizes and kills depends on the prey. Prey animals that have the potential to injure the cat or are somewhat noxious to the cat are treated differently
from relatively innocuous or easily overcome prey. For example, a small cat may
toss a small animal about before delivering a killing bite, but a large cat grasping a
large prey animal could easily be gored by a horn or receive a debilitating kick from
a sharp hoof if the cat does not deliver a killing bite immediately and effectively.
There is no room for error when the prey presents a danger, and even a rabbit can
deliver a kick that can debilitate a lynx.

Seizing the Prey

With certain prey, a cat must learn always to employ special tactics. Normally a cat can see its prey, but it may also use auditory triangulation by moving its ears in a way to estimate position. Servals and sand cats have exceptional auditory perception in the ranges that their rodent prey communicate, and servals can locate and pounce on prey that they probably cannot see (see *How Do Cats Hear?*). A rule among carnivores is that if prey is not visible, go for the head, where a bite is most likely to kill it quickly. When they cannot see their prey, primitive nocturnal carnivores kill by rushing toward the noise and biting, but probably without the efficiency that cats have achieved. Cats have evolved complicated stalking procedures to approach prey they can see but cannot close in on directly without scaring it away.

The only risk to a stalking cat is that, being focused on the prey, it too becomes vulnerable to predation. A cat uses cover to remain undetected while approaching a prey animal as well as to protect the cat's own blind side during the killing sequence. The seizing and killing stage is risky in other ways too, though. A cat may be somewhat ambivalent about going after a prey animal that it recognizes as dangerous; it may approach and retreat before actually attempting a kill. If the prey animal is completely novel to the cat and perceived as dangerous, the cat may ignore it or even flee. In most circumstances, though, if a cat is familiar with the prey or the prey is small and flees, the cat will attack and try to kill it.

For a large cat (a tiger, for example) to kill a large prey animal, the cat must first grab the prey or bring it down before delivering a killing bite to the nape or throat. When the tiger is very large and the ungulate small, the cat may simply knock the prey down with a forepaw and bite its nape or throat. In other situations, it may seize the prey by the neck with just the teeth, or seize the neck with the teeth and forepaws simultaneously, or seize with the forepaws and then grab the neck with the teeth. A tiger may also make a series of moves that take advantage of the prey's own momentum by reaching over the prey animal's back to bite the throat. Chuck McDougal and John watched tigers use a series of moves they call counter rolling, in which cats grabbed their large ungulate prey by the neck and dragged it in the opposite direction from which it had fallen, rolling it over onto its opposite side so that the hooves pointed away from the cat when it made its killing bite to the throat (Seidensticker and McDougal 1993).

The Killing Bite

The canine teeth are a cat's all-important killing tool. Cat biologist Paul Leyhausen believed that the small cats he studied delivered a very precise killing bite to the prey's neck vertebrae: "A cat's canine teeth strike the cervical vertebrae and the tooth then inserts itself between the vertebrae like a wedge, forces them apart, and thus severs the spinal cord partially or completely. . . . The canine teeth are exceptionally well suited to forcing things apart but certainly not to biting firmly with their tip on something very hard" (1979, 33). He suggested that this bite, guided by feel, is both a very effective killing technique and also protects the canines from potential damage caused by the crushing contact with bone. Further, he thought there was a "lock-and-key" relationship between the diameter of a cat's canines and the size of the cervical vertebrae in its preferred prey.

The flexible killing behaviors of tigers, however, indicate that a close match is not necessary between canine diameter in cats and the size of the cervical vertebrae in prey. Tigers have the largest canines and jaw lengths of any felid and kill the largest prey. Chuck McDougal and John investigated killing bites by dissecting the necks of the ungulate prey of tigers. In some, they found that the cervical vertebrae immediately behind the skull in fact had been crushed by the tiger's canine teeth, and the vertebral column was severed in many cases. But besides the killing bites that were directed toward the nape, others were directed toward the side of the neck and the throat. Tigers also kill by strangulating prey without puncturing the spinal column, the trachea, or the jugular. The size of the neck of very large prey animals precludes the tiger's canine teeth from striking the cervical vertebrae, so strangling such prey is the only way the tiger can kill them. Young tigers learning to hunt also first strangle their prey, only later adding the nape bite to their repertoire.

ARE CATS EVER HURT BY PREY?

Hunting is a risky job, particularly for the large cats killing big dangerous mammals bearing sharp hooves, horns, tusks, or antlers. One Canadian study reported that 27 percent of natural mortality among pumas was related to injuries sustained while trying to capture prey. Pumas have been found at the base of a cliff with a broken skull and with a piece of mountain mahogany branch piercing the brain case. In Idaho, wildlife biologist Maurice Hornocker interpreted the story of an encounter between an adult female puma and an elk by "reading" the tracks and marks they left in the snow (Hornocker 1970). The puma stalked the elk, and the two slid down a sharp incline, whereupon they hit a tree. The elk escaped. When Hornocker captured the puma three weeks later, he found her jaw had been broken, both lower canines had been torn out, and she had puncture wounds in her shoulder and hip.

When a gemsbok, a large African antelope, was shot, the partly decomposed body of a leopard was impaled on its long horns. Tigers are often gored by water buffalo and gaur. Porcupine quills also take their toll on inexperienced cats. At 15 kilograms, these large, slow-moving rodents can provide a sizable meal. However, many

Servals often hunt small rodents they can hear but not see. They rush toward the prey, then grab and chomp down on it with their teeth. This is a safe and effective way for cats to dispatch small prey.

Cheetahs kill large prey by holding the victim's throat between their teeth until the animal suffocates. This same technique is used by other cats, including tigers, snow leopards, Eurasian lynx, and caracals, when they are attacking prey much larger than they are.

This leopard has killed an impala with a bite to its nape. The cat's canines either crush a prey animal's cervical vertebrae or sever its spinal column. The killing nape bite is employed by all cats, but this killing technique is the most difficult for a young cat to master.

Lions risk lethal injury from the flying hooves of wildebeests and other ungulate prey. Hunting large animals is dangerous, and cats are often hurt by swift kicks or goring antlers and horns. Even small prey can inflict debilitating wounds. Leopards and tigers sometimes end up with face full of porcupine quills when they try to kill this spiny rodent.

leopards and tigers that turn to man-eating have been previously debilitated by porcupine quills; pumas have also been found with porcupine quills in their face. These cats eventually learn to flip porcupines over onto their back before they kill them, but even then, quills are routinely found in their scats in some areas.

WHY DO CATS SOMETIMES KILL MORE THAN THEY CAN EAT?

From time to time we hear reports of a puma or leopard killing many prey animals within an enclosure, usually livestock such as sheep or goats, and then eating only one, if any, before the cat is driven off. This is greatly upsetting to the livestock owner, and, if possible, the cat is hunted down and killed. This "surplus killing" makes little biological sense: it has little survival value, and it even seems to be a waste of effort on the predator's part. A prudent predator should use its food resources efficiently.

Why kill more than can be eaten? Generally in mammals, a feedback mechanism prevents further eating once an animal is full. Cats show very little activity after a substantial meal and may not hunt for hours or even days, depending on the species. If a cat finds itself in close quarters, inside a livestock holding area, for example, with many potential prey animals, the search and detection part of its hunting behaviors is no longer necessary, and somehow the seizing and killing parts click into a loop without any controlling feedback through consumption and satiation. In this situation, killing does not necessarily lead to eating; instead, the predator does some more killing. This behavior is adaptive to this particular situation, because normally if a cat is able to kill several prey quickly, it will do so and eat the carcasses later. But, outside a livestock pen, this is rarely possible.

HOW MUCH PREY DO CATS KILL AND HOW OFTEN?

On average, big cats consume 34 to 43 grams per day per kilogram of their weight. Puma-sized cats consume about 56 grams per day per kilogram, and ocelot-sized cats consume 60 to 90 grams per day per kilogram. The very small cats (less that 2 kilograms) consume 125 to 150 grams per day per kilogram of their weight. Yes, smaller cats must eat proportionally much more than larger cats. On the basis of per gram of cat, the metabolic rate of a black-footed cat or other very small cats is much greater than the metabolic rate of a tiger; thus, the black-footed cat needs relatively more fuel to stay alive (see *What Does Size Have to Do with It?*).

How does this translate into killing frequency? In the dry South African shrub, Sliwa (1974) found that black-footed cats killed a bird or mammal every 50 minutes, on average. Geertsema (1985) observed servals in Tanzania's Ngorongoro Crater capturing 5,700 to 6,100 prey per year, equal to about 3,950 rodents, 260 snakes, and 130 birds. In Doñana National Park, a single Iberian lynx killed about one rabbit a day; a female of this species with two kittens needed to kill three rabbits a day (Delibes et al. 2000). A Canada lynx was found to kill about one hare every 2 days, for 153 hares per lynx per year (Nellis and Keith 1968). The distance traveled per kill depended on the number of hares taken. Lynx killed about the same number of hares whether hare numbers were low or high but had to travel farther each night to do so when the numbers were low (Okarma et al. 1997). Subadult and female Eurasian lynx without kittens in Poland's Białowieza forest killed 43 red deer or roe deer per year, and adult males killed 76. In Wyoming, Anderson and Lindzey (2003) observed pumas to kill on average a large mammal every 7 days, but this ranged from 5.4 days per kill for a family to 9.5 days for a single subadult, for an average annual total of 52 large mammals, which included mule and white-tailed deer, elk, pronghorn antelope, moose, coyote, and livestock.

John tracked a female tiger with two 6-month-old cubs living in Nepal's Royal Chitwan National Park, where preferred tiger prey, such as wild pigs and sambar, are abundant. This female killed every 5 to 6 days, which amounts to 60 to 70 kills per year (Seidensticker 1976). Further studies in Chitwan have revealed that, on average, female tigers make a kill once every 8 to 8.5 days, or roughly 40 to 50 kills a year. Studies of lions suggest a slightly lower average. At an estimated average requirement of 5 kilograms of meat per day to stay alive, each lion will account for about 30 medium-sized kills per year.

DO CATS STORE FOOD?

Scientists refer to animals' food-storage behavior as caching. Many cats cache the remains of kills that provide more food than they can eat at one time, but it may be a challenge for a cat to keep scavengers from stealing the meal. Spoilage may also be a problem, especially in hot weather, whereas a carcass may freeze in cold weather and be preserved. Pumas drag a large kill to a secluded site, bury the carcass with snow, leaves, grass, dirt, or other debris, and stay nearby until the meat is finished, usually after about three nights. Pumas drape themselves over a frozen carcass to defrost the meat. Most cats cache their leftovers but do not necessarily guard them. Both tigers and pumas may travel to a resting site several kilometers from the cache and then return. When a tiger returns to its meal, it does so with caution and stealth: if a pig or a brown bear is scavenging the kill, the tiger has a chance to kill another big meal. The habit of returning to an unguarded kill, however, leaves tigers vulnerable to a bigger threat than losing dinner: people intent on killing tigers lace cached carcasses with poison.

When many scavengers are present, leopards drag a large kill up into a tree, draping the body over a limb. This keeps thieves (more formally referred to as kleptoparasites) such as lions, spotted hyenas, and ground-feeding vultures at bay. However, this kind of caching also has its risks. In a Namibian study, leopards had potentially dangerous interactions with other carnivores at 12 percent of their kills. Leopards dragged all carcasses, except hares and birds, on average 140 meters into thick underbrush, where they then fed; they dragged their kills into trees only 3 percent of the time. A kill in a tree alerted lions, and the lions then pursued and tried to kill the leopards. Where cover is scarce, such as in the Kalahari Desert of southern Africa, leopards dragged prey as far as 700 meters.

In contrast, cheetahs do not cache food or return to carcasses. Unable to fend off the large scavengers of the African plains, cheetahs are often driven off their kills before they are finished eating, so they just have to hunt again.

Smaller cats cache uneaten prey too. Black-footed cats regularly cache prey, ranging from insects and birds to shrews, rodents, hares, and scavenged antelope fawns, by covering it with soil and grass. Frequently, these cats leave carcass parts sticking out, which might help them find the spot again when they return within a few hours. In the Białowieza Primeval Forest in Poland, Eurasian lynx were often found to move the carcasses of their ungulate kills to dense vegetation or push them under fallen logs, where they covered them with soil, leaves, moss, deer hair, or snow to protect them from scavengers so that they could feed on the cached prey for several days. Roe deer were dragged farther than the heavier red deer, but both usually less than 100 meters. In this study, 80 percent of the cached lynx kills were subsequently fed upon by scavengers, most often wild pigs.

To avoid scavengers, most cats drag or carry their prey to sheltered places to eat and cover the remains of an unfinished kill to preserve it for another meal. Leopards sometimes haul large prey like this warthog up into a tree, out of the reach of nonclimbing lions and hyenas.

HOW DO CAT NUMBERS AFFECT PREY NUMBERS AND VICE VERSA?

The dynamics of the relationship between predator and prey vary considerably from ecosystem to ecosystem. Biologists call it a top-down process when predators influence the number of prey; when prey influence the number of predators it is called a bottom-up process.

As a rule of thumb (although with some variation), about 10 percent of the sun's energy is fixed into green plants through photosynthesis, about 10 percent of that energy passes to herbivores, and about 10 percent of the herbivore energy is available to carnivores. The primary productivity, or amount of plant growth, in an ecosystem is influenced by its amount of light, warmth, moisture, and nutrients, all of which differ considerably from place to place and over the seasons. This bottom-up effect determines the amount of plant food that is available to the herbivores, its primary consumers, in a particular place and time. The number of herbivores is limited by the quality of the plant food, the form it is in, and the defenses the plants have evolved to keep from being eaten. For example, tropical rain forests are some of the most productive places on Earth, but most of the plant material is in the form of tree trunks, which most herbivores can't eat; the plant material in leaves is out of reach to terrestrial herbivores. Also, leaves deploy defenses, such as toxic or noxious secondary compounds, to protect them from being totally eaten by insects. These same compounds reduce the quality of the leaves as food for arboreal herbivores, such as primates and sloths. The actual amount of plant growth that is digestible and available determines how many herbivores can live in a place, and this, in turn, determines the number of predators, or secondary consumers, that can live there. The plants are subjected to top-down forces when the herbivores eat them; herbivores are subjected to these same forces when the carnivores eat them.

Some ecologists argue that predators regulate the number of herbivores, which in turn limits the damage herbivores do to vegetation. Other ecologists argue that the bottom-up forces are most important in determining how ecosystems work. Most conservation biologists agree that both top-down and bottom-up forces are important in most ecosystems. Over the last 75 years, biologists have addressed this question in countless studies, most of which have focused on the impact of the predator on its prey, many fewer on the effect of prey on their predators.

In many ecosystems, the density of wild cats is positively correlated with the density of their primary prey. More snow leopards occur where there are more blue sheep. More lions occur in areas where prey numbers are highest year around. Cheetah and leopard densities are strongly correlated with "lean-season prey biomass"—the combined weight of all potential prey available when food for prey is least abundant. Going a step further, biologists Chris Carbone and John Gittleman found that, across the whole range of carnivore sizes from mongooses to polar bears, about 10,000 kilograms of available prey biomass are required to support 90 kilograms of any given species of carnivore. This means that 10,000 kilograms of prey biomass are required for every 0.5 tiger, 2 leopards, or 45 rusty-spotted cats. Ultimately, however, predator populations are constrained by the reproductive rates of their prey rather than by the amount of prey present at any one time (see *What Does Size Have to Do with It?*).

Carnivore populations adjust to prevailing food supplies through changes in reproductive output, survival of offspring, and dispersal of young away from their natal areas. A study in Canada showed that as snowshoe hares declined, lynx body fat declined and the initial reproduction of female lynx occurred at a later age, pregnancy rates declined, and kittens were less likely to survive. A study in Tanzania demonstrated that the survival of lion cubs is also directly related to lean-season biomass of prey; cubs' survival rates during their first year of life were more than twice as high where prey numbers were high.

Adult mortality also increases when food is short. Cats often starve, and hunger also leads cats into conflict with people. During very hard winters in the Russian

Far East, when deer and wild pigs starve and die, tiger-human conflicts increase, as they do wherever hunters have killed most of the tiger's prey. Food shortages also lead to more strife among individuals of a species because they trespass into one another's territories looking for food. Moreover, the size of territories is influenced by the amount of food available.

When the rabbit and hare populations declined in the shrub-covered, semidesert of Idaho, the size of bobcat home ranges increased fivefold; that of lynx increased threefold during a snowshoe hare decline in the Yukon. Lynx also become nomadic, disperse farther, and travel farther each day when hare numbers are low. In Africa, the size of the home areas of lion prides is correlated with lean-season prey availability. During periods of prey scarcity, the number of transient or dispersing lions increases, something also seen in feral domestic cats. Dispersing individuals normally do not reproduce and have higher rates of mortality than nondispersing individuals have.

Understanding cause and effect in the predator-prey relationship is challenging, both because secretive wild cats are very difficult to count accurately and because defining what is prey for a cat can be difficult. In his study of pumas, John found that numerous deer and elk resided within the home areas of pumas, but only those prey adjacent to adequate stalking habitat were available to the cats (Seidensticker et al. 1973). Pronghorn antelope normally live in the open plains and shrub steppes of the American West, where adults, with their exquisite eyesight and great speed, are nearly immune to predation. But when pronghorns enter rugged, bushy terrain in central Arizona, they become vulnerable to pumas, and puma predation on female pronghorns has been high enough to decrease the size of the population.

Cats may also take prey species that have become newly available to them. Biologists Joel Berger and J. D. Wehausen (1991) described a great example. Pumas were historically absent from the Great Basin Desert of Nevada, but when livestock grazing changed the vegetation from grass to shrubs, the number of shrub-eating mule deer increased. Pumas followed the mule deer. Until then, the bighorn sheep living there had experienced little predation, but with the arrival of the pumas they were decimated. Similarly in the Sierra Nevada, after pumas arrived, their numbers had to be controlled to protect a remnant endangered population of bighorn sheep. Studies of introduced species, specifically feral cats introduced to islands, offer abundant evidence of the profound effect a predator can have on "naïve," vulnerable prey. Island breeding seabirds are particularly vulnerable to predation, because they lack any behavioral or morphological defenses against it (see *Why Are Feral Cats—and Pet Cats—Sometimes a Menace?*).

When several predator species coexist in an area, competition complicates the predator-prey relationship, or at least our understanding of it. Tigers and leopards seem to depress sambar and Indian muntjac numbers in India's Nagarohole National Park. But in comparing Nagarohole with other Asian tropical forests, John and his colleagues Charles McDougal, Ullas Karanth, and Mel Sunquist found that the occurrence and relative densities of tigers and leopards depended largely on the relative densities of different size-classes of prey (Seidensticker and McDougal 1993; Karanth and Sundquist 2000). Leopards are more abundant where medium-sized prey are abundant for them and where large prey are abundant for tigers. Where medium-sized prey are few, tigers exclude leopards, because leopards can't get at the large prey. But where neither large nor medium-sized prey are abundant, leopards do better than tigers, because leopards can supplement their diet with prey too small for a tiger to survive on.

Ecologists Charles Krebs, Tony Sinclair, and associates found that predators control prey populations when predators have a direct density-dependent effect, which they often do (Krebs et al. 1995). Density dependence is the correlation between prey density and predator-caused mortality, measured as a percentage of the prey populations. When prey populations are low—"in the pit," as ecologists say—predators can affect their numbers; but if something happens to reduce predator numbers, prey can break out of the pit. This leads to prey "swamping" predators: prey are so numerous that the size of their populations is unaffected by predators taking their requisite share. This phenomenon has been seen in many prey-predator systems. For example, even the combined predation of cheetahs, lions, wild dogs, and spotted hyenas has little effect on the migratory herds of ungulates in the Serengeti. In this system, the large ungulates are regulated by their food supply, not by predators.

In synchrony across Canada, snowshoe hares exhibit 8- to 11-year population fluctuations, during which their numbers vary 10- to 25-fold. Is it lynx predation or food availability for hares that drives this cycle? Is it bottom-up or top-down? Lynx take more hares when the prey are abundant, and the number of lynx fluctuates 2- to 10-fold during the hare cycles. Biologists once thought that plant food shortages occurring when hare numbers were greatest set the decline in motion, because the hares began to starve, having eaten themselves out of house and home. But subsequently their predators caused even higher hare mortality rates, until hare numbers became so low that lynx began to starve and die, giving the hares a chance to rebound. Extensive experimental work has failed, however, to demonstrate that food shortages initiate the decline and has showed instead that the cycle is most likely due to interactions of all three factors—food plants, hares, and predators—throughout the cycle.

Recently biologists Ricardo Moreno and Jacalyn Giacalone set out to determine the impact of ocelots on Barro Colorado Island, a tropical rain forest island in Panama managed by the Smithsonian Institution as a research station (Leigh

2002). Scat analyses revealed that an ocelot's annual take was 22 two-toed sloths, 18 three-toed sloths, 15 white-faced monkeys, and 18 agoutis (medium-sized rodents). A 10-kilogram ocelot eats about 33 times its own weight a year; therefore, as much as 10 metric tons of mammal prey a year were eaten by the 30 or more ocelots living on the island. The entire island is thought to support 68 metric tons of mammals.

In studying tropical forest islands of various sizes that were created when a river was dammed in Venezuela, ecologist John Terborgh and his associates found that where predators of vertebrates were absent, the densities of rodents, howler monkeys, iguanas, and leaf-cutter ants were 10 to 100 times greater than on the nearby mainland, where mammalian predators, including ocelots and as many as four other species of cats, were present (Terborgh et al. 2001). This strongly suggests that predators regulate these prey populations. Further, without predators to control the herbivores on these islands, the densities of seedlings and saplings of canopy trees were severely reduced. These biologists concluded that predation pressure on herbivores "has been weak over much of the earth since the eradication of megafauna by stone age hunters, so bottom-up regulation has become widespread, creating aberrations that have spawned the top-down versus bottom-up controversy" (p. 1926).

WHY DO MOST WILD CATS LIVE ALONE AND WHY ARE SOME SPECIES EXCEPTIONAL?

Most carnivores live solitarily, and most wild cats are no exception. This style of living suggests that, among carnivores, conditions rarely favor group living outside the reproductive period. In other words, the costs of group living outweigh the benefits. Group living carries several disadvantages: it increases the chance of being detected by a predator or competitor; it increases the risk of disease and parasite transmission; and it increases the chances of aggression and injury, which a solitary-hunting felid can ill afford. So living at low densities is characteristic of species at the top of the trophic pyramid, or food chain. Being obligate carnivores predisposes most cats to living alone.

Only 10 to 15 percent of all the carnivore species live in groups outside the breeding season. Among the cats, feral domestic cats live in groups around very rich resource bases, such as docks, dumps, and farm sites, but live solitarily when their food resources are less abundant and well dispersed. Cats don't even live in mated pairs, which is the basic social unit in canids.

In the pair-forming canids, the males provision the females and the young by bringing food back to the den; they either carry it back in their mouth or they regurgitate food they have eaten. Cats do not have the capacity to bolt down large chunks of meat or eat prey whole, nor do cats seem to have the capacity to greatly distend their stomachs, as wolves do, for example. Cats slice up their prey carefully with their carnassials. When they kill larger animals, they conceal them by thick cover. A prey item that cannot be eaten in a single meal is cached, and the cat returns for several meals. John radio-tracked a female leopard in Nepal that stayed away from her cubs for days at a time while she hunted for sambar deer, killed one, and consumed it, one meal at a time, until it was finished. She spent her time guarding her prey rather than returning to her 6-week-old young, which she had stashed in a stand of tall thick grass no more than an hour's travel time away.

Most cats live alone, probably because most hunt more efficiently as individuals. With the exception of cheetahs and lions, cats stalk and ambush prey alone. Even in situations where large prey are abundant, as occurs in some habitats of tigers, lynx, pumas, and jaguars, these species do not form groups to take advantage of some of the potential benefits of sociality. Perhaps their prey is just too difficult to catch and kill in the thick cover where they hunt. Unlike cats, canids cannot bring down a large prey animal alone without great difficulty, and their chasing style of hunting lends itself to cooperation, as we see in wolves. Also, most of the smaller canids are not strictly meat-eaters, supplementing their diets with fruits and vegetable matter, which may allow them to forage and feed communally, something not seen in the cats.

With the exception of lions and free-ranging domestic cats, cats seem to be generally intolerant of conspecifics (members of the same species). This leads to females living spaced apart in home ranges, which may either overlap to some degree or be mutually exclusive of one another in what are referred to as territories. Generally, biologists have thought that the young, upon reaching nearly adult size, either leave their mother's home range on their own, or are abandoned, or are even sometimes ousted by their mother from her territory. However, we are learning that in many cat species, even though young males leave their mother's territory, not all female offspring do. If the mother's territory has proven to be adequate to support her and her family and even more resources have become available with the departure of young males, a young female will take over part of her mother's area rather than leave it, thus narrowing the mother's original range. This may happen with several different female offspring until, eventually, the mother finds that she cannot survive in the small area remaining, and she abandons it. In effect, she is squeezed out by her daughters or, in some cases, her granddaughters, and this marks the end of her reproductive life.

Except in lions and free-ranging domestic cats, though, female cats and their female young don't stick together. We do not know if artificial selection during domestication led to greater tolerance among domestic cats, but it has never been

Barely visible in the grass, this leopard exemplifies the secretive, solitary lives of most cats. A mother and her young are the largest group formed by all species except lions, cheetahs, and some feral domestic cats. Once they leave their families, males spend the rest of their lives alone, only occasionally meeting a female to mate.

naturally selected for in wild cats. Intolerance toward other cats of the same species is hard-wired, so that most wild cats cannot live in groups even if the resources are sufficient to support it. So far, only group-living domestics have been studied in resource-rich urban environments. But recent studies of fishing cats and other wild cats living around resource-rich Asian cities and towns will help to answer this question.

Among lions, both males and females live in groups; these coalitions of males and prides of lionesses live where food is abundant in the form of ungulates. Also the presence of many other lions is important for reproductive opportunities. Male lion coalitions are more successful in gaining breeding access to pride-living females by working as a unit to defend a pride when they are resident or to oust the current resident males; some very successful male lion coalitions gain access to more than one pride of females by displacing additional males. Female lions can live in groups because of their hunting style: related females hunt cooperatively to take large prey in open habitats. Large carcasses last a long time and are hard to defend, so it is easier and genetically beneficial to share with a relative rather than lose food to unrelated lions. The further advantage of group living in female lions is that they can join together to defend their cubs against infanticidal males.

Male cheetahs also live socially, in coalitions of two or several males. As a unit, these males are more successful in holding down large territories containing important resources that are attractive to breeding females, such as food and den sites. Female cheetahs, on the other hand, live and raise their young alone. As soon as the young are mobile, they and their mother begin to follow herds of migrating ungulates, leaving the males and their territories behind.

So in lions and in cheetahs, high densities and localized females create intense competition among males but favor the formation of their alliances (often involving related males), which can win in battles against single males or smaller coalitions. Living in groups also allows more complex social behavior to evolve, such as cooperative hunting, joint maternal care, organized vigilance, and protection from intruding conspecifics.

HOW DOES A CAT'S HOME RANGE RELATE TO ITS REPRODUCTIVE SUCCESS?

With no pair bonds among felids and no direct contribution among males to the care of their offspring, how do males improve their reproductive success? They can do so by mating with more than one female. Males attempt to include as many females as possible within their territories. Ultimately, however, males too are restricted by their energy needs and by the amount of space they can actively and effectively defend. The resulting spatial pattern results in what biologists call intrasexual territoriality. Female cats are *contractionists*, and males are *expansionists*. A female contracts the space she uses to include just the area that contains the resources needed for herself and her offspring. Males *expand* the space they use to the extent possible while maintaining the ability to defend it, thereby retaining exclusive breeding access to all the females living within.

John studied pumas' use of space in different seasons and habitats (Seidensticker et al. 1973). He found that male territories are always significantly larger than those of females and their young living in the same general areas as the males. Pumas have young during all seasons of the year, so there is never a nonbreeding season. In pumas and in other cats, males kill the young of a female they do not know, so that she enters estrus again and he can sire his own offspring with her. This is a strong reason for males to maintain their territories year around to prevent the intrusion of other adult breeding males.

Sometimes both males and females make forays outside their usual territories, males to gain access to other females, females to mate with males living farther away. These can be very dangerous excursions because of the risks of encountering a resident territorial cat. However, there are also advantages. Both male and female cats are known to change territories if an adjacent, better territory becomes open through the death of the resident. Females are much more conservative than males in shifting home ranges, because they know their areas and where and when they can predictably find the prey they need to kill to support their litters. By moving, they lose this advantage, at least in the short term. Males are on the lookout for the area richest in females that they can defend with the least effort and danger from other males.

HOW BIG ARE CAT HOME RANGES?

A home range or home area is the space an animal uses that contains all the resources it needs to survive, including food and water, shelter from the elements as well as places to sleep and hide from predators, mates, and places to rear young. Cats also make occasional forays outside the area they usually cover in order to check out developments in the social and physical neighborhood. The knowledge a cat gains in doing this may be essential to its survival or to improving its reproduction potential, even though these forays are not included when we calculate the size of the cat's home range.

Home areas may overlap among individuals. If there is no overlap, the home range is called a territory; a territory is occupied more or less exclusively by a single individual or a group, as in a pride of lions, and is maintained by means of repulsion through overt defense or advertisement. Usually, males defend their territories from other males, and females defend theirs from other females. Among cat species, home ranges are most often territories. A home range of either type provides advantages: residents know from experience where and when to find the resources they need to survive, where potential mates can be found, and where the special places they need to rear young are located. Newly independent young cats as well as old cats that have been displaced from their territories may wander through the territories of established residents, which puts them at risk of attack. Or these displaced individuals may live in very marginal habitats, where they await their chance to take a resident's place, either through the death of the resident or through combat.

Cats, like most carnivores, respond to fluctuating resource supplies in different ways. Where the primary prey migrate between distant summer and winter areas, a cat either moves from one part of its territory to another to follow the food, or it expands and contracts its use of the territory to encompass the expanding and contracting range of the prey. This is a flexible strategy. In tropical environments, where resources are more constant and equally distributed throughout the year, cats tend to use their entire territory year around. If the resource base varies from season to season and from year to year in the same areas, the cat has two options: adjust the territory size to the resource base needed to support itself (the flexible strategy) or set territorial boundaries at an invariant, usually large, size that fills the cat's needs at all times (what Torbjorn von Schantz [1984] calls the obstinate strategy).

Similarly, just as female cats tend to be contractionists and males expansionists, some entire species are contractionists, occupying the smallest economically defensible territory possible. Others species are expansionists, having territories with resources well in excess of minimal requirements for survival. Cats use both of these strategies, depending on the species and the environment.

Compare, for example, the size of tiger territory in the prey-rich tropical riparian forest and tall grass floodplains in Nepal with that of tigers living in the prey-poor temperate, oak-pine forests of the Russian Far East. In Nepal, the tiger's prey doesn't move seasonally, although the densities of prey vary from one habitat type to another. In Russia, the pigs and deer, the mainstay of the tiger's diet, shift between winter and summer ranges. In Nepal, female territories were found to be 16 to 20 square kilometers, and those of males were 60 to 72 square kilometers. In Russia, females used areas that ranged in size from 245 to 415 square kilometers in summer; winter areas were smaller and overlapped in varying degrees with the summer-use areas.

In California, female puma home range sizes in the northern Coast Range have been found to average 90 square kilometers in summer and 100 square kilometers in winter; males' average 300 square kilometers in summer and 350 square kilometers in winter. In the dry Sierra Nevada, where the puma's prey, mule deer, shift between winter and summer ranges, puma home range size for females and males,

Cats regularly patrol the boundaries of their large territories to fend off intruders of the same sex looking for territories of their own. With territories ranging from about 20 square kilometers to nearly 300, depending on the habitat and the sex of the animal, tigers must cover a lot of ground to defend their home turf.

respectively, is 541 and 723 square kilometers in summer and 349 and 469 square kilometers in winter.

In contrast, the home range of the tiny kodkod averaged 2.69 square kilometers in one southern Chile site. At another site, where the biologists studied kodkod movements in both forest and farm land, males ranged over 0.7 to 17.4 square kilometers and females over 0.6 to 1.46 square kilometers.

Cheetahs living in the Serengeti exhibit huge differences in territory and range size between the sexes. Male coalitions hold territories with dens and nonmigratory prey, which attract females raising young until the young are old enough to follow migratory ungulate herds. Female cheetahs do not defend their home ranges and thus are not considered territorial. The average size of a female home range was found to be 833 square kilometers, and that of male territories was 37.4 square kilometers.

In general, larger mammals usually have larger home range sizes. Strict carnivores, such as the cats, have larger home ranges than similar-sized omnivores and herbivores. However, home range sizes vary among the cat species, among individuals of the same species and the same sex occupying different areas, and between the sexes within the same species, as our examples above demonstrate. The variables that affect home range size include habitat preference of the cat, habitat preferences of the prey, prey density, prey distribution in the area, and competition among carnivore species and among members of the same cat species.

In a study in California that compared 57 individual home ranges of pumas, sex, body mass, relative deer abundance (their primary prey), and season all influenced home range size, although not in a linear way. For instance, males always had larger home ranges than females, even when the two sexes were about the same size. The home ranges of young adults were larger than those of older adults. Surprisingly, deer abundance was a weak indicator of home range size, which was also true of the reproductive status of a female and body size in either sex. Other factors thought to influence home range size included the density of other pumas in the area, the densities of alternative prey species, and the density of roads and human developments.

Where John studied pumas in Idaho, the size of home areas also was not directly related to the density of their primary prey in winter, mule deer and elk. Of course, a puma's home area had to support enough deer and elk for the cat to survive, but beyond that parameter home area size was set by topographic conditions. The most important element was abundant places where these cats could successfully stalk and kill their prey. Home ranges of both males and females were smallest where the canyons were very rugged and broken. Where the canyons turned into basins with large treeless areas, puma home areas were larger. In both types of habitat, 45-kilogram female pumas were killing 360-kilogram male elk, so the pumas had to use every possible terrain advantage they could.

HOW DO SOLITARY CATS COMMUNICATE?

Cats generally live and hunt alone but spend their lives embedded in dynamic societies. These societies include their neighbors, their mates, offspring, and parents, as well as strangers that enter the scene from time to time. Solitary-living females (such as tigers) may interact face to face and share their range with one litter of young (male and female) at a time, whereas group-living females (lions) may share their range with their female young from several litters and continue to do so when those female offspring are adults.

Cats are not truly asocial; it's just that many of their social interactions are conducted through long-distance communication, with loud vocalizations and scent marking (see *What Is Scent Marking?*). At the same time, they have a rich repertoire of short-range communication behaviors, including purring in many species, vocalizations of friendly greetings, and cheek rubbing between courting males and females, as well as facial expressions and body postures that are often seen in aggressive encounters.

An aggressive cat on the offensive growls and rotates its ears forward, stands with its back parallel to the ground or with its weight shifted to its front legs, and lashes its tail from side to side. At the other extreme, a submissive or defensive cat hisses, bares its teeth, flattens its ears, and may slink away, crouch down, or even roll over onto its back. When a cat responds to aggression with submissive behavior, it is essentially "crying uncle." By giving up to a clearly stronger opponent, the cat can live to fight another day.

WHAT IS SCENT MARKING?

All cats leave scent marks in the environment; these reveal information about an individual's sex, reproductive status, identity, occupation of a particular area, and perhaps its age, dominance, and health status. Rates of scent marking are highest at territorial boundaries, when these boundaries shift or when the entire territory changes in possession, and when females are approaching estrus. The odorous substances involved in this marking are in the urine, feces, and saliva, and are also produced by special glands on the tail, chin, lips, cheeks, and feet. Scent marking also has a highly visual and spatial component. Marks are not deposited at random. Instead, marking is most frequent at boundaries between territories, along well-used paths, at intersections of paths, and on particularly prominent habitat features, such as a large, conspicuous tree that stands a bit apart from others. Tigers are even selective about the species of tree on which they spray.

To draw even more attention to spots that have been scent-marked, cats rake their claws across the bark of trees, leaving visible scars as well as odors from the glands on their paws. Many cats scrape the ground with their hind feet, leaving a bare patch to highlight a spot on which they've urinated. Some cats deposit feces on scrapes, either accompanied by urine or alone. Ocelots, bobcats, and other wild cats repeatedly defecate in the same "toilet" spot to create a highly visible, smelly mound. Sand cats, which live in deserts with little or no vegetation, create piles of feces on heaps of sand. Geoffroy's cats are unusual in depositing their feces in the crooks of tree branches as high as 3 to 5 meters above the ground, a surprising behavior for a cat that spends most of its time on the ground.

Pumas do not spray, but males make carefully placed scrapes. In his study of pumas, John found that scrapes were 15 to 46 centimeters long, 15 to 30 centimeters wide, and 3 to 5 centimeters deep (Seidensticker et al. 1973). Most were placed from 0.3 to 2 meters from a fir or pine tree or from a rock face or overhang. On slopes, scrapes were always made on the downhill side of trees. Scrapes were rarely located on trails; usually they were off to the side. He found scrapes near the mouths of canyons, in draws, and on ridges, and they appeared to be located where the lay of the land indicated easy passage.

Solitary cats communicate by leaving smelly messages on objects in their environment. Other cats read the messages to find out about the animals that left them, learning their sex, reproductive status, and how recently they have been in the neighborhood. This bobcat is leaving such a message by rubbing its cheek gland, which produces an odorous secretion, across a fallen tree.

Cats want others to find their scent messages—communicating at a distance is safer than potentially dangerous face-to-face encounters. As many cats do, this ocelot is raking its claws along a tree trunk, leaving deep, highly visible gouges as well as scent from glands on its paws. This makes it easy for other ocelots to find the message.

These visual and spatial signals of scent marking help to ensure that the marks are not missed by their intended audiences. Both male and female territory holders, for instance, prefer to avoid the risks of a fight with a transient and thus want the transient to find and be deterred by marks that delineate the boundaries of their territory. Similarly, a female approaching estrus wants to advertise her status so that potential mates will seek her out, and before mating, a male may also increase his rates of marking. A male black-footed cat sprayed 585 times on the night before mating, whereas the more usual rate for a resident male is 10 to 12 times an hour during active times of the day.

Scent marks, unlike vocal or visual signals, persist in time and thus are uniquely suited to communicating about the past. A fresh scent mark reveals that it was laid down very recently and the cat that made it is very likely nearby. An old, faded mark says the opposite. Thus a transient male, for instance, may be undeterred from entering another male's area if the marks are old. Cats may also make occasional forays into the territories of others and leave scent marks, as if to test the competition. This means that scent marking is an important job and must be performed repeatedly and frequently. A few studies suggest that scent marks persist for 3 to 4 weeks before they are ignored as signals of ownership.

Cats also overmark. A male that comes across the scent mark of another will mark over it, in a sort of game of one-upmanship. Cats occupying adjacent territories concentrate their marks at the borders, keeping others informed that they are still in residence.

Cats criss-cross their territories regularly both to keep track of what's going on in their area by reading the smelly messages left by their neighbors and strangers and to leave messages of their own. In animals in which any face-to-face confrontation has the potential to escalate to deadly aggression, conducting their social life at a distance is the safest way to live.

DO ALL CATS SPRAY?

Spraying and claw raking are the least endearing traits of domestic cats kept indoors, but this natural behavior is difficult to eliminate. One of the most common scent-marking behaviors of cats is spraying urine, which has been observed in all of the cat species that have been studied except for the puma. As cats travel, they stop often to spray, usually on vertical objects such as tree trunks or shrubs. The cat backs up to the object, raises its tail, and sprays urine backward. In most species, both males and females spray, but males tend to do so more often, and the cat penis is designed to let males direct urine at just the right height above ground: at nose level, so that the mark is in easy reach of its audience. Some cats have been observed to spray at very high rates. In a study of small wild cats in captivity, female Canada lynx, sand cats, Asian golden cats, and servals sprayed on average seven or eight times an hour. In the wild, male servals have been observed to spray about 40 times per kilometer traveled, or about 46 times an hour, and females slightly less often, 15 times per kilometer and about 20 times an hour.

DO ALL CATS BURY THEIR FECES?

There are situations in which cats refrain from advertising themselves through scent marks. In some species, such as leopards, transient individuals without territories do not mark, preferring to move about undetected. Also, females with young may stop marking and bury their feces when depositing them near the den. This behavior has been observed in bobcats and black-footed cats, perhaps to keep infan-

Like all cats except pumas, cheetahs back up and spray urine on trees and other vertical objects. Performed by both sexes, but more frequently by males than females, urine spraying leaves scent messages that are very important in territorial defense.

ticidal males, as well as potential predators, from using odor clues to find the young. (Similarly, among pumas, only 2 scrapes of every 100 that John found were likely those of females, and these females were without kittens. Males were responsible for all of the other scrapes.) Female pumas with and without kittens cover their feces with piles of pine needles, as the kittens themselves do. But unlike domestic cats, wild cats do not always bury their feces and more often leave them conspicuously. Even the wildcats (*Felis silvestris*), from which the domestics arose, leave their feces in prominent locations where they will be found by others. Farm-living domestic cats bury their feces only on their home area and leave feces and urine on display elsewhere. It's also been suggested that very small cat species are more likely to hide their feces to avoid being detected by predators. Pallas' cats reportedly bury their feces, and it is possible that when Geoffroy's cats deposit their feces high up in trees, they do so as a form of burial, in that the feces are hidden from nonclimbing predators.

In a friendly gesture, a female lion rubs her head on another's. Such rubbing applies one cat's scents to the other and is seen in all cats. Two cats smelling like each other may enhance their familiarity and reduce tension.

DO ALL CATS HEAD-RUB?

Cats rub their head and many different parts of their body against objects in the environment and against one another. They also rub their head, cheeks, chin, and neck against urine and other marks left by other cats and even roll their entire bodies on the ground or in urine, rotten meat, and other strong-smelling substances. Cheek rubbing between cats, most often in a breeding pair greeting each other after a separation, is a friendly gesture. In rubbing, cats both apply their scents and pick up other scents from their environment and from their social partners. When our pet domestic cats rub against us, they are doing the same thing. In making their odor more like that of their environment, and vice versa, cats may be camouflaging their own scent, which may make it more difficult for prey to smell their approach. Exchanging odors with social partners, either directly or indirectly by rubbing in their urine marks, may enhance their relationship: familiar, similar odors may reduce tension when they meet face to face.

WHAT IS FLEHMEN?

Flehmen is a behavior mostly performed by males, but also sometimes by females. It occurs when males sniff a urine scent mark or the genital region of a female. It is

This male lion is displaying a behavior called flehmen, which is a grimacing lip-curl that brings chemical signals called pheromones into contact with the vomeronasal organ in the roof of the mouth. This organ detects chemicals whose molecules are too large to be detected by the olfactory receptors inside nostrils. Pheromones convey information about a cat's reproductive status.

also seen when a cat sniffs any novel odor. When flehming, a cat curls its upper lip and appears to be sneering or grimacing. At the same time, it stands very still and breathes slowly with a faraway look in its eyes, as if concentrating very hard. What it is doing, however, is exposing the two openings of the vomeronasal organ in the roof of the mouth, behind the front teeth, to better receive pheromones—the chemicals that send messages to the parts of the brain concerned with sexual behavior—and using its tongue to carry drops of whatever it is sniffing into contact with the vomeronasal organ. Pheromones are heavy molecules that cannot be picked up, or "smelled," by the receptors that line the inside of the nostrils. Flehmen is probably universal in cats and is widespread among mammals. Most mammals—

except marine mammals and, controversially, humans—possess vomeronasal organs that are important in detecting pheromones related to reproductive behavior.

WHAT ARE SOME CAT VOCALIZATIONS?

German biologist Gustav Peters has devoted his life to studying the vocalizations of cats, and much of what follows is based on his work (Peters 1984, 2002; Peters and Hast 1994). All cats share a basic set of acoustic signals, with one or two vocalizations unique to one or more species. Except for purring and hissing, vocalizations are produced by oscillations of the vocal folds during exhalation; purring occurs during both exhalation and inhalation, and hissing may not involve the vocal tract at all. All cats spit, hiss, growl, and snarl during hostile interactions. The first three in the sequence indicate increasingly aggressive motivation, to the point where continued growling reveals a cat's readiness to attack. The snarl, on the other hand, is a defensive sound; in some species, a submissive cat being attacked may shriek.

In friendly, close-contact interactions, cats produce noisy, low-intensity calls. In most species, this is emitted as a gurgle—pulsed sounds that, in some species, Peters compares to cooing in pigeons and, in others, to the sound of bubbling water. In lions and leopards, the equivalent call is a puff, which sounds like a series of stifled sneezes; in tigers, jaguars, snow leopards, and clouded leopards, it is a prusten, which is like the snorting of a horse.

All cats mew as well. A mew can range from a soft sound used by females in close contact with kittens all the way to a very loud roar used for long-distance signaling. In many species, the mew sounds similar to the familiar meow of domestic cats, but the comparable call in pumas sounds like a whistle, and in the big cats it includes grunts. Roaring is discussed in more detail below.

Other vocalizations have been heard in a few species. Pumas, and a handful of other cats, use a "wah-wah" call when two animals approach each other; Eurasian lynx and domestic cats chatter by clacking their jaws when prey is close but out of reach. Cheetahs chirp in several different situations (see *Why Do Cats Roar, Chirp, or Meow?*).

DO BIG CATS ROAR AND SMALL CATS PURR?

Until fairly recently, it was a common assumption that the cats could be divided into the big cats that roar but do not purr and the small cats that purr but cannot roar, a distinction believed to be rooted in differences between the hyoid structures in these groups. However, careful analysis by Peters and Hast (1994) has revealed that it isn't necessarily so. In five species—lions, tigers, leopards, jaguars, and snow leopards—the hyoid contains a cartilaginous ligament; in the rest of the cat species the hyoid is all bone. Hence, many people believed that the cartilaginous ligament was necessary for roaring but prevented purring in the big cats, and vice versa in the small cats. This distinction also had taxonomic significance, because it was formerly used as one of the bases for dividing most of the felid family into two subfamilies: the Pantherinae and the Felinae (a third subfamily was composed only of the cheetah). However, modern analysis of both vocalizations and genetics refutes this. First, snow leopards do not roar, and, surprisingly, neither do tigers. In fact, what we generally call roaring in tigers is a sequence of calls whose acoustic properties are different from those involved in the roars of lions, leopards, and jaguars, which share a grunt call lacking in tigers' "roars." Neither do clouded leopards, now placed in the Panthera Lineage with the big cats, roar.

The anatomy of the vocal system of the roaring cats (including tigers) reveals how they make these very loud calls. According to M. H. Hast, "The entire vocal mechanism of the roaring *Panthera* . . . is analogous to the brass trumpet" (1989, 120). The vocal folds in the larynx are the mouthpiece, connected to the vocal tract—a straight tube that can be adjusted in length (formed of the supraglottal larynx and pharynx)—which at the other end is connected to the cat's wide open mouth, analogous to the bell of the trumpet. Combining a high-energy sound generator (the vocal folds) with an efficient sound radiator (the vocal tract ending in the open mouth) enables a cat to "use its vocal instrument literally to blow its own horn with a 'roar'" (p. 120).

Whether the big cats purr remains an open question, however. We know with certainty that 15 of the 40 cat species purr; there are no data on the presence or absence of purring in another 9 species. And about the remainder, including all of the Panthera Lineage species, scientists are unsure whether they purr or not, although Peters believes that all but the Panthera Lineage cats will be found to purr. As curator of mammals at the Smithsonian's National Zoological Park, John observed and listened to lions and tigers at very close range for many years; he never heard either species purr. So all we can say with confidence is that many small cats purr and none are known to roar, whereas big cats roar and may or may not also purr.

WHY DO CATS ROAR, CHIRP, OR MEOW?

We've seen few things surprise visitors to the National Zoo as much as a chirping cheetah does. When first hearing this sound, they look for a noisy bird and are amazed when they realize that it's actually a noisy cheetah. A cheetah's chirp, a lion's roar, and a domestic cat's meow, despite the differences in how they sound to us, function as long-distance signals to other members of their species. All cat species have a vocalization for long-distance signaling. Although how various species use this signal is not well known, available evidence indicates that females use it to attract mates before and during their estrous periods and that both sexes may use it to maintain contact with desirable companions and to proclaim possession of an area so that conspecific competitors stay away.

Cheetahs chirp in three different natural situations. Estrous females chirp to attract males. Both sexes chirp when afraid or in distress. Males chirp when separated from members of their coalition and also chirp upon their reunion; females and young chirp when separated and reunited as well. In a study conducted at the National Zoo, we found that the chirps of different individuals were quite distinct, suggesting that cheetahs may be able to recognize one another by their chirps alone. What's more, chirps have an acoustic structure with abrupt changes in frequency and amplitude, and they are emitted in a series. This is believed to facilitate locating the source of a call and may thus help separated individuals find one another.

Jon Grinnell and his colleagues have explored when and why lions roar in Tanzania's Serengeti National Park (Grinnell 1997; Grinnell and McComb 2001). A roar actually consists of three different sounds. It begins with soft, low moans, progresses to loud, high-energy roar sounds, and concludes with a series of staccato grunts. Grinnell believes the grunts make the sound easy to locate (both males and females roar to maintain contact), and the roar sounds perhaps convey information about the individual's sex, condition, status, and identity.

Male lions roar to proclaim their possession of a territory, but only while they are resident in a pride of females that they are prepared to defend from other males. Outside their territories, resident males are silent, as are nomadic males (those without territories). In situations in which males have nothing to defend, silence is golden. Roaring would only attract the attention of resident males, which would try to find and oust the intruders. The one exceptional situation, when a single nomadic male joins a single female in estrus, he roars to display his possession of the female.

Females also roar to advertise their ownership of a territory, even though there is some risk in this. Nonresident males are attracted to roaring females, but these are exactly the males that female lions want to avoid. Male lions that take over a pride of females immediately set about killing any cubs that can't get away or their mothers can't defend. The larger the number of females in a pride, the better able they are to defend their young from marauding male newcomers. Males looking for a pride seem to know this: they are much more likely to approach a single roaring female than a trio of roaring females.

On the other hand, Grinnell found that estrous females roar in order to attract males and also employ some trickery to incite competition between male coalitions.

A larger coalition of males can usually best a smaller one, so it's to a female's advantage to be associated with a large coalition. A female sometimes tests the waters by sneaking off to mate with males in a nomadic coalition. Believing they are now territory owners by virtue of being in possession of a female, these males begin to roar, only to face the rage of the female's resident males, which rush to defend their territory from the interlopers! If the new males are in a larger coalition, they may win the battle and oust the resident males; if not, the nomadic males are driven away. In either case, the female and her pride end up with the better male coalition. Females may have to employ such ruses, because male lions are loath to fight unless there's a good chance of winning. One way to gauge their chances is simply by assessing the size of the potential opposition, and male lions make this assessment by listening to the other side's roars.

The tiger's long-distance call, although technically not a roar, is usually called a roar nonetheless (see *Do Big Cats Roar and Small Cats Purr?*). Their roaring is emitted in various situations, including by a female in estrus to attract a mate, by a male to announce his arrival, and by a female to call her young. Tigers may also roar after making a kill. It is very likely that roaring proclaims territorial possession, as it does in lions. If the acoustic qualities of an individual's roar change with the animal's size or health, another tiger may be able to tell if it's worth the risk of fighting to take over a territory. But there's more to a tiger's roar than meets our ears. Scientists recently discovered that tigers are able to produce infrasonic sounds—very-low-frequency, deep sounds that humans cannot hear. The roar includes such sounds, which are notable for their ability to travel long distances even through dense vegetation, which tends to obscure higher-frequency sounds. This may help explain how tigers can stay in contact while occupying very large territories. Although we cannot hear these sounds, we may feel them, just as we can feel the pulsing bass of very loud rock music. Heard suddenly at close range, a tiger's roar knocks you back like the blast of a speeding train.

WHY DO CATS PURR?

Along with the cat's meow, purring is the most familiar of felid vocalizations. People are so familiar with this sound that it is used in metaphors—the engine of a well-tuned car purrs, for instance. Purring conveys contentment; it says "all is well." Purring is a close-range vocalization, heard most often between a mother and her young or in other close-contact friendly contexts, including between a person and his or her contented domestic cat. Purring is unusual in that it can go on for minutes—it is more like a song than a shout—and continues while the cat both inhales and exhales. Very young kittens purr, even without interrupting nursing. Purring is generally a low-intensity soft sound, which may keep predators from detecting and following the sound to the den. People can hear domestic cats purring from at most 3 meters away, but the purring of larger cats generally carries farther than that of smaller cats. Because cats usually purr when they are in body contact, Gustav Peters believes that "the body surface vibration during purring may serve as a tactile signal in addition to the auditory one and may even be at least as significant in conveying the senders' message" (2002, 264).

Among the cats known to purr are the cheetah, puma, and jaguarundi in the Puma Lineage; the serval; the ocelot, margay, and oncilla in the Leopardus Lineage; the bobcat, Eurasian lynx, and probably Canada and Iberian lynxes in the Lynx Lineage; the caracal in the Caracal Lineage; the wildcat and black-footed cat in the Felis Lineage; the Asian golden cat in the Bay Cat Lineage; the marbled cat; and the leopard cat in the Leopard Cat Lineage.

HOW OFTEN DO CATS REPRODUCE?

Approached from the point of view of adult females, cat reproduction is easily summarized. Adult females of the larger cat species—lions, tigers, jaguars, leopards, snow leopards, pumas, and cheetahs—produce litters every other year; smaller cats do so about annually or, rarely, twice a year. The exceptions are some of the small South American cats—ocelots, margays, and oncillas—which sometimes skip a year, and domestic cats, which can produce as many as three litters in a year. Thus, in an idealized scenario, a female lion or tiger, for example, mates with a male, produces a litter of young 100 or so days later, and resumes sexual activity nearly 2 years after that, when her young become independent and she is able to care for new ones. This summary, however, hides a great deal of variation, and females rarely reproduce like clockwork.

Whether the young of any particular litter survive to independence influences the following birth interval, and, as we describe below (in *How Long Do Cats Live?*), infant mortality can be quite high. In cats that are not strictly seasonal breeders (and many species are not) females will mate soon after losing a litter. Male lions take advantage of this by killing the young in a pride they've newly taken over, so that the mothers will breed with the newcomers more quickly, within a few days or weeks. This increases the chances that the new males' offspring will be born and mature before they are ousted by another male group. In these cases, the birth interval will be less than the average 2 years.

Other factors act to lengthen the birth interval. Females do not conceive in every period of estrus. In one well-studied population of pumas, fewer than one-quarter of breeding associations between a female and a male or males resulted in kittens being born 92 or 93 days later. Studies of tigers and lions suggest that the chances of conception in any one estrus in these species are only 20 to 40 percent, and only 2 of 13 courtship associations (15 percent) observed in leopards in South Africa resulted in births. Female lions mating with the same males that fathered their previous litters take fewer estrous periods to conceive than females mating with new males.

Many female cats will continue to come into estrus until they conceive. Estrous cycle length varies considerably, even among some closely related species. The lion estrous cycle averages 16 days, while that of leopards is 46 days, and jaguars 37. There is also considerable individual variation within some species. The estrous cycle of tigers has been reported as both 15 to 20 days and 40 to 50 days, for instance.

Canada lynx are seasonal breeders, with mating occurring during about a month in the spring, so that babies are born in the warm months of May and June and have time to grow fairly large before the weather turns harsh. Females exhibit only one estrous cycle per year, and those that fail to conceive or lose their litters will not mate again until the next year. Moreover, the reproductive success of Canada lynx corresponds with changes in the abundance of hares, which form the mainstay of their diet. When hare numbers crash, few or no females produce live young in the first year of the low numbers, and few young are born—and few of these survive—in the second year. During hare abundance, however, pregnancy and birth rates range from 73 to 100 percent for adults females, with 50 to 83 percent of young surviving.

Snow leopards live in habitats with harsh winters as well, and nearly all births occur in the late spring, with more than half in May alone. Estrus appears to be very brief, and a female has at most only one estrous cycle each year. Females that lose litters seldom mate again until the following winter, while females raising young will not become receptive to mating until the second winter. Similarly, Amur tigers, which inhabit the northern temperate forests of the Russian Far East and northeast China, tend to give birth in the spring.

Tropical cats are generally less seasonal in their breeding, and, in many species, females can become pregnant in any month of the year. However, births may cluster before peaks in the availability of prey to help ensure that mothers have plenty to eat to support the high-energy demands of lactation. Servals, for instance, are born about a month before the peak breeding season of their rodent prey. South African leopard births peak during the birth peak of impala, in the early part of the wet season. Not only are prey more available then; they are also more easily caught when wet-season vegetation gives the cats better stalking cover. Young leopards, too, are better concealed from predators when they are hidden among the vegetation.

How many times a female cat reproduces in her lifetime is not known with certainty for any species. One exceptional female tiger in Nepal's Royal Chitwan National Park lived for 15 years, 10 of them in the same territory. She produced five known litters of cubs; her first was born in 1975, her last in 1985, so she actually did reproduce, on average, about every 2 years for 10 years. Eleven of her 16 total young survived to independence. Female lions produce their first litters at 3 years of age, give birth on average every 2 years, and for most of their life, litter size is constant, averaging 2.5. At 14 years, however, average litter size drops to 1, and by 17, if they survive so long, they no longer reproduce. According to lion biologist Craig Packer, "Maternal survival is only important during the first year of a cub's life, so when a lioness reaches the age of fourteen . . . she can expect to live another 1.8 years—long enough to raise the last cub born" (1998, 26). Thus, although a female lion could potentially raise 15 or 16 young in her lifetime, she rarely would. Sunquist and Sunquist (2002) report that one Serengeti lion had seven litters in 12 years but raised only two of those litters, for a total of six cubs.

Female cheetahs live particularly short lives in the Serengeti. A 25-year-long study revealed that the females that survived to independence lived on average just over 6 years (at a maximum of 13 years) and produced fewer than two cubs that survived to independence. Among the 108 females in the study group, 8 females raised between 7 and 10 cubs to independence, and 55 raised none. Females gave birth for the first time at about 2.4 years of age and experienced average birth intervals of 20 months when they raised cubs to independence. Females that lost cubs swiftly mated again, sometimes as early as 2 to 5 days after the loss.

HOW DO CATS FIND MATES?

Except for lions, in which males and females share a territory and interactions are frequent, male and female cats seldom meet face to face, although individuals living in the same neighborhood know one another. Male territories generally overlap those of more than one female, but they avoid using the same areas at the same time. Their mutual avoidance is believed to be mediated by scent marks (see *How Do Solitary Cats Communicate?* and *What Is Scent Marking?*). Both males and females deposit scent marks. They do this primarily by regularly spraying urine on trees and bushes, making scrapes of bare soil on which they may also urinate or defecate, and depositing feces in prominent locations as they travel through their territories and patrol their territorial boundaries.

These scent marks are like signs: a fresh marks says, "I'm working here now; you find somewhere else." A stale sign says, "I've moved along; it's safe for you to come in." As a female approaches estrus, however, the message changes. Changes in the chemical composition of the urine correlated with the rise in estrogens that precedes full estrus and ovulation are read by males as "Come find me; I'm ready to mate." To ensure that the local male gets the message, a female increases the rate at which she scent-marks. The male, in turn, routinely patrols all areas of his territory to check on the status of the females living there.

Other behaviors performed more frequently by female cats in estrus include restlessness, rolling, cheek rubbing, body rubbing, and vocalizing. These behaviors among female domestic cats in estrus are familiar to many people and may even be directed at their human caretakers. Wild cats observed in estrus appear to behave similarly, although most of their behavior has been observed in captivity.

Females approaching estrus and in it may also use their species-specific, long-range vocalization to call in males. Female tigers, for instance, sometimes roar but not always. Sunquist and Sunquist (2002) describe a female tiger that roared during five consecutive estrous periods, some days roaring day and night, as many as 69 times in 15 minutes, and then roared not at all during the next estrous periods they observed.

Jon Grinnell recounted an episode he observed while studying lions in the Serengeti:

> As part of my research I tranquilized a number of male lions to take blood samples and to attach radio collars to aid in relocating them. Once, I immobilized a male in consort with an estrous female. I hadn't seen the two mating before this, but after the male was fast asleep the female decided it was time, and walked sinuously by his sleeping form, back and forth, her tail lashing her receptive scent under his nose. When after a few minutes of this he still hadn't responded, she walked off—one can only guess at her thoughts—and started roaring, even though it was midday and a highly unusual time for a lion to roar. She continued roaring until the male awoke from his drugged sleep and rejoined her. Why did she roar? Probably the simplest explanation is that since she was in estrus and her first male had lost all interest in her, she meant to attract a new one. (Grinnell 1997, 12)

Roaring and other long-distance vocalizations may attract not only the resident male but also other males looking for a chance to mate. Transient males moving through the area but avoiding another male's territory may miss a female's scent-mark advertisement but still hear her call. Possibly females vocalize in order to attract other males to the scene, to create a "may the best man win" competition among them (see *Why Do Cats Roar, Chirp, or Meow?*). Females in many mammal species, including elephant seals, right whales, and bison, appear to incite competition to ensure that the highest-quality male available fathers her young. Breeding encounters have been observed in the wild in only a few species, but in tigers, jaguars, pumas, cheetahs, wildcats, caracals, and others, estrous females are often attended by more than one male.

A female domestic cat may mate with multiple males that attend her without much fighting among themselves. As in lions, these males are usually relatives, which may reduce the competitive urge. But dominant males likely achieve most of the matings. Male lions in a coalition do not have equal access to a female in estrus; one male often monopolizes the female for much of her estrous period, which lasts 4 days or so. Only when his interest wanes will other males get a chance to copulate.

HOW DO CATS MATE?

Once a male and a female cat have found one another, a period of careful, high-tension courtship follows. When two such predators get together, there is always the chance that a false move on either's part will escalate into lethal combat rather than copulations. The female is initially coy, aggressively rebuffing the male's attempts to get close and mount her, even though she may have solicited his advances by walking sinuously by him, rolling in front of him, and touching him. The male meekly gives way, only to try again and again until gradually the female becomes fully receptive and accepts his advances. When the female permits, the pair may lick and rub each other. Finally, the female adopts a crouching posture, called lordosis, which makes her accessible to the male. He then mounts and they copulate. The male may or may not hold the female's neck between his teeth during mounting and copulation, but immediately after ejaculation, the female twists her body to dislodge him, often swatting at him with her paw. Intromission or ejaculation may stimulate a copulatory vocalization by male or female or both. Female tigers, for instance, yowl after ejaculation, as male snow leopards and both male and female pumas do.

Once the male and female begin copulating, they do so with reckless abandon in many species. Individual copulations are generally brief, lasting from a few seconds to less than a minute, although caracals are reported to engage in long copulations ranging from 90 seconds to 8 or 10 minutes. Copulations are usually numerous, however. Tigers may copulate tens or hundreds of times during an average 7-day estrous period. In one study, puma pairs were observed to copulate 2 to 20 times per day over 6 to 11 days, and in another study, a pair copulated 23 times in 10 hours. Captive jaguar pairs have been observed to mate 100 times a day over 7 days. A captive lion pair copulated 360 times in 8 days, and lions in the wild have been observed to copulate every 20 minutes or so over several days, during which time males don't even bother to eat. In captive snow leopards, copulations lasted from

A male and female lion mate repeatedly, as often as every 20 minutes for the several days of the female's estrus. Other big cats copulate at similarly high rates, whereas small cats tend to be more restrained, perhaps because of their greater risk of a predator taking advantage of their distraction. Cheetahs, which are heavily preyed on by lions, are almost never seen mating.

15 to 45 seconds and were repeated 12 to 36 times a day for 3 to 6 days. Captive female black-footed cats exhibited very short periods of receptivity—just 5 to 10 hours—but pairs still copulated as many as a dozen times during that time. Ocelots copulate just 5 to 10 times a day for about 5 days. Cheetahs are highly secretive breeders, and mating is rarely seen in the wild. The estrous period is short, just 2 or 3 days, and brief copulations usually occur at long intervals and at night. Big cats have the luxury of enjoying lives free from the threat of predators. This may account for their flamboyant matings over many days compared with the more discreet pairings of cheetahs and the small cats for which there are data.

Cats are believed to be induced ovulators, but this has not been demonstrated for most species. Induced ovulation means that the stimulation of copulation is required for a female to ovulate, unlike, for example, human females, who spontaneously ovulate about every 28 days even if no male is present. Induced ovulation may be an advantage in species in which males and females live widely separated from one another. Waiting to ovulate until copulation actually occurs ensures that opportunities for successful matings are not missed. This idea is supported both by the frequency of mating among female cats and by the special attribute of the penis of males, which is covered with small, stimulating spines. However, it appears that induced ovulation is not a hard-and-fast rule in cats. Lions, tigers, and leopards may be spontaneous ovulators, at least under some captive conditions, and spontaneous ovulation is the norm in bobcats, Canada lynx, and some fishing cats. On the other hand, multiple copulations may increase the probability of conception: a captive female tiger that copulated 100 times during one estrus conceived but did not do so during an estrus in which she copulated only 30 times.

HOW LONG IS GESTATION?

Gestation length in cats ranges from 100 to 114 days in lions and 100 to 108 days in tigers to 63 to 68 days in black-footed cats and 66 to 70 days in rusty-spotted cats. Most of the variation is a straightforward function of size; that is, the larger the cat, the longer the gestation length. Remaining variation is related to the total weight of the litter, so among cats of similar size, those with heavier litters have longer gestation periods than those with lighter litters. Cats in the South American Leopardus Lineage are unusual in this regard, however. Female ocelots, margays, oncillas, and the rest often bear a single young but have longer gestations than other cats their size with larger litters. Three-kilogram margays, for instance, usually have one (at most two) young after a gestation of 76 to 84 days, while 3-kilogram Pallas' cats average three or four young (the range is from one to six) after a gestation of 66 to 75 days. In general, mammals with lower basal metabolic rates have longer gestations than mammals of similar size with higher basal rates. This may account for the longer gestation in margays, which have relatively low basal metabolic rates for a felid. However, jaguarundis also have relatively low basal rates, but their gestation lasts about as long as is expected for a cat its size. Gestation length is also longer in species with young that are relatively more precocial versus those with young that are more altricial (that is, young with a relatively high degree of independent activity at birth versus young that are born very immature and helpless). Margay young are relatively large at birth, like those of oncillas. Oncilla young weigh between 92 and 134 grams at birth, while similar-sized rusty-spotted cats and black-footed cats bear young between 66 and 70 grams and 63 and 68 grams, respectively.

HOW MANY BABIES ARE IN A LITTER?

Litter sizes in cats range from one to ten, with ten reported only in domestic cats and seven or eight in lions, tigers, cheetahs, lynxes, and a few other species.

Like this female nursing five youngsters, cheetahs generally have large litter sizes, perhaps to compensate for very high rates of infant mortality in this species. Although litters of six or seven or more occur occasionally, most cats typically have litters of two or three young, and ocelots, margays, and oncillas usually have just one.

Females rarely, if ever, raise that many young. Typical litter sizes for most species are two or three, although one is most frequent in ocelots, margays, and oncillas. Cheetahs have relatively large litters at birth, averaging 3.5 in the Serengeti, a number that may be an underestimate. Large litters may be an adaptation to the extremely high rates of mortality suffered by young cheetahs, primarily due to lion predation. Mortality before cubs emerge from their dens usually involves the entire litter, but after emergence, only some of the litter may be taken. The larger the litter size, therefore, the greater the chances that some of a female's young will survive. Correlated with large litter size are low newborn and litter birth weights for the body size of the species and high growth rates of the young.

Female cats invest little in their young until they are born, perhaps because the demands of hunting preclude females' carrying very large babies. All cats have small, semi-altricial young, although there is variation among species. Newborn cats' eyes are closed. They can move about but not very well. They have fur, but their ability to maintain their body temperature is poor, and they rely on their mothers entirely for food and protection.

WHERE DO CATS HAVE THEIR BABIES?

Female cats seek out a variety of sheltered sites, usually called dens or lairs, to give birth. They do not build dens but instead use existing natural features, which may be no more than a space hollowed out of dense vegetation or a hollow tree, an abandoned burrow of another species, or a crevice among rocks. Even sociable female lions secretively withdraw from their pride mates to give birth and hide their cubs in dense thickets or rocky outcrops for the first few weeks of their life. Ideally, dens offer some degree of safety from predators, and they shelter the young from severe weather. After giving birth, mothers may stay with the young constantly for a few days but must soon leave the cubs tucked away in the den in order to resume hunting.

Nestor Fernandez and his colleagues (2000) performed a detailed analysis of den use in Iberian lynx, which use two types of dens. Birth dens are invariably in natural hollow trunks of either cork oak or ash trees. The trees chosen were wider, had more dead branches, and more holes in the trunk than nearby trees of the same species, and were usually surrounded by high bushes. Fernandez believes the key advantage of hollow tree dens is that the mother's body heat can maintain the den at a fairly even temperature compared with the fluctuating external ambient temperature. This could be critical to keeping the vulnerable young warm in their first few days, until their own ability to thermoregulate improves. Mothers begin to stay in the natal den area about a week before giving birth. They leave the young in the birth den for 2 to 3 weeks, then move them to bush dens when they start becoming mobile, because the hollow tree dens leave little room for the young to move about. Bush dens are in large, thick bushes (usually lentisk) that offer lots of hiding places in and around them, including rabbit warrens to scoot into and tangled stalks, brambles, and bark and cork. A female may move her cubs to five different bush dens in sequence, but she gets less fussy about selecting a den as the cubs grow older and are better able to protect themselves. Females may change dens so that accumulating odors and other signs of their presence don't alert predators; prolonged used of a single site might also lead to larger populations of ectoparasites. Disturbance by other animals or humans triggers a move as well. After the cubs are about 2 months old, they begin to leave the den.

Bobcats are similar to Iberian lynx in their den use. Natal dens are most often in natural rocky areas and caves, but other sites include abandoned beaver lodges and

These caracal kittens are tucked into a lair formed of thick brush. Mother cats give birth in such secluded, relatively safe spots, then leave their cubs or kittens there while they go out to hunt. Mothers usually move their young to several new lairs before they are old enough to travel along with the mothers.

storage sheds. The same natal dens may be used for many years. They too move young up to five times to other dens with progressively less protection than the natal dens.

One factor that influences den site selection in some species is plentiful nearby prey, because lactating females have large energy requirements, which would be best fulfilled by finding prey close to home. This did not seem to influence den site selection in Iberian lynx (den site areas had low numbers of rabbits), but this may be because the territories of these lynx were quite small, so they didn't have very far to go from any point within them.

The location of a female's lair relative to prey has been shown to be important in cheetahs as well. Cheetahs are most often born in lairs in patches of thick, tall vegetation, in marshes, or in rocky outcrops, where they remain for about 8 weeks. Females far from prey were more likely to abandon their cubs. One Serengeti female's lair was 12 kilometers from the local concentration of Thomson's gazelles; she ate far less than other mothers, and after spending 3 days trying to capture gazelles 9 kilometers from the her lair, she gave up and abandoned the cubs.

Black-footed cats are born and raised in abandoned aardvark and porcupine burrows and termite mounds, which adults use in any case for shelter during the day. Mothers change dens several times. Where aardvark burrows are available, leopards may also use them as den sites.

Female cats' insistence on privacy during parturition, which usually occurs at night, and during the early days of maternal care has kept us from knowing much about the related behavior of wild cats outside captivity, so most of what we know is based on observations of domestic cats. The events surrounding parturition in cats are fairly typical of most mammals. Females about to give birth may be restless, and they lick their nipples and genital area clean. Once babies are born, the mother licks them clean of the birth membranes and cuts the umbilical cords with her teeth. Lying with the young keeps them warm, and they soon begin to nurse. Mothers vocalize softly to their young and show them much affection.

DO FATHERS HELP RAISE YOUNG?

The only role that a father cat plays in caring for his young is to defend his territory so that another male doesn't move in and kill them. Infanticide is best known in lions, but it occurs in many cat species, including tigers, ocelots, Canada lynx,

Nursing of cubs and all other parental duties are strictly the responsibility of females in lions and other cats. In the social lions, males may play with, or at least tolerate, young that they may have fathered, but generally the only parental role males play is keeping at bay other males, which may kill young not their own.

pumas, and feral domestic cats. A female tiger completely avoids all males while she is rearing cubs and may cede a kill to a male rather than expose her cubs to him.

Other than defending the territory, caring for the young is solely the mother's job in all species of cats except lions. Unlike wolves and some other canids, female cats raise their young without the assistance of a mate or other helpers, and the energetic costs of feeding young cats are very high. A nursing female ocelot, for instance, was observed to double her normal activity levels and spent 17 hours a day hunting during lactation. This likely contributes to the small litter sizes and relatively slow growth rates seen in most cats. Only under conditions of very high prey abundance can a mother feed both herself and more than a few needy cubs.

Lions are exceptional. Females in a pride often give birth at the same time, and they may rear their young communally in crèche—the lion equivalent of a day-care center. Cubs nurse from all the mothers if they can, but mothers try to nurse only their own cubs. Large cubs may do better in this system than small ones, however, because they can compete better for access to nipples.

Resident male lions tolerate cubs fathered by males in their coalition but no others. When cubs are old enough to feed at kills, adult males eat first, giving their cubs no special treatment, and when prey is scarce they keep cubs from eating at all.

HOW LARGE ARE NEWBORN CATS AND HOW FAST DO THEY GROW UP?

Across all mammals, newborn babies generally weigh between 1 and 20 percent of the average adult female weight. But, given the variation in litter sizes, a better comparison can be made using the relationship between the weight of an entire litter and the average adult female weight. Among carnivores, the range is very large, with total litter weights ranging from 0.1 to 22 percent of female weight, with bears being the lightest at birth. Among cats, the range is usually between 2 and 6 percent, heavy compared with bears but light compared with canids, whose litters are 7 to 16 percent of female weight.

Also, compared with other carnivores, large cats grow relatively slowly, and small cats have intermediate growth rates; canids, often supported by two or more adults, grow quickly. Being a mother cat is extremely hard work, and she does it with no help. Domestic cats with a large litter at the peak of its demands for milk have 2.5 to 3 times the normal energy requirements. One study of pumas predicted how many kilograms of deer per day a female would have to kill to raise from one to four cubs. A single adult female needs about 2 kilograms of deer per day, whereas a female and a single cub need about 6 kilograms per day by the time the cub is able to feed itself, and a female and four cubs need almost 16 kilograms. Other data show that a mature female puma kills a mule deer about every 16 days, and that number increases to one every 10 days for a female with two 2-month-old cubs and to one every 3 days for a female with three 15-month-old cubs. Put another way, a female alone can get by on one or two black-tailed jackrabbits a day, but a female and three 15-month-old cubs need seven or eight jackrabbits a day. Female servals with kittens spend twice as much time hunting and travel longer distances than normal to do so.

Like measures of other reproductive parameters, the time from birth to adulthood varies with body size, with small cats growing up faster than big cats. There are several ways to measure when a cat is "grown up." One is the age at which it becomes sexually mature, defined as being capable of successful reproduction. However, animals may be capable of reproducing before they actually do so. Males, especially, may be prevented from breeding because they cannot wrest a territory from older, larger males. Among females, access to a territory and their nutritional plane may affect when they first breed. Females often take a part of their mother's territory or a territory nearby—or, in the case of lions, remain on the same territory and share it with the mother and aunts. In some species, such as bobcats, this means that a female might mate as a yearling, especially when food is abundant. Yearlings tend to be far less successful, however, than 2-year-old females.

Other measures are age at dispersal, when young adults or near-adults leave their mothers, and age when they reach adult body size. Among the larger cats, offspring generally leave their mothers at about 2 years of age, whereas, among the smaller ones, young leave at 1 year or earlier. Dispersal age may be younger than age at sexual maturity, though, and varies among individuals, perhaps depending on when the mother has her next litter of cubs. Tigers disperse at between 18 and 28 months or later but do not achieve adult size for several more years. Males grow faster than females from the beginning. Males are sexually mature at 3 or 4 years of age, and females at about 3 years, but most do not give birth until they are 3.5 or 4 years of age. At the other end of the size spectrum, female black-footed cats are sexually mature at 8 to 12 months, and males even earlier, although their chances of actually breeding so young in the wild are unknown. Domestic cats are sexually mature as early as 7 months of age.

Just as their gestation length is relatively longer and their litter size smaller than other cats, the growth rates of the small South American cats in the Leopardus Lineage are relatively slow. Ocelots grow more slowly relative to body size than any other cat, not achieving adult size until they are 2 to 2.5 years of age; breeding begins at about the same time. Oncillas are weaned at about 3 months of age, are near-adult size at just under a year, and are sexually mature at 2 to 2.5 years. Margays reach adult size early, at about 1 year—it is generally advantageous for arboreal species to grow up quickly to cope with the challenges of foraging in trees—but they do not breed until they are 2 or 3.

Day-to-day, week-to-week growth and development are very well documented for the domestic cat, and the major changes in behavior and physiology follow a similar pattern in all cats that have been studied, albeit with much individual variation within species as well as variation in timing among species. For instance, kittens in a small litter may grow more quickly than those born in a large litter, and cubs in larger-bodied species generally begin to practice hunting and follow their mothers to kills later than kittens in smaller species.

Born with their eyes closed and little able to move or keep themselves warm, kittens of domestic cats quickly begin to grow and acquire skills. Their eyes open at 1 to 2 weeks, and at 2 weeks they achieve several major milestones: they hear and orient to sounds, their deciduous (baby, or milk) teeth begin to erupt, they begin to take their first tentative, wobbly walks, and they can follow a visual cue. At 3 to 4 weeks, they begin to thermoregulate, they can find their mother using visual cues, they can smell as well as an adult and hear nearly as well, they begin to play, and they eat their first solid food; at 4 weeks, they begin to run. From 5 to 7 weeks, they become very nearly mini-adults. They can now practice predatory behavior and are able to kill mice, their sleep pattern and ability to thermoregulate become adult-like, and they move like adults. By the end of 7 weeks, they are weaned from mother's milk. Finally, at 10 to 11 weeks their coordination is fully developed, and at about 12 weeks, adult teeth begin to replace the baby teeth. In the next months, they will continue to grow and perfect their social and hunting skills until they are adults, at about 10 months.

The time of achieving certain milestones differs little among species. For instance, tiger babies' eyes open at about 11 days (varying from 0 to 20 days), lions at about 6 days (varying from 0 to 11 days), leopard cats at about 10 days (5 to 15), bobcats at about 6 days (3 to 9), and even the slow ocelots are not too far behind, with eyes opening at 15 to 18 days. Similarly, the range of variation reported for young cats first eating solids is not very wide, with wildcats doing so at 28 days, leopard cats at 44 days, Eurasian lynx at 50 days, cheetahs at 33 days, leopards at 42 days, and tigers and lions at 56 days. Other milestones are more variable, usually according to the species' body size. Tiger mothers, for instance, begin taking cubs to large prey kills or bringing small prey to their cubs when they are 3 to 4 months old, compared with domestic cats, which do this when kittens are 4 weeks of age. Similarly, tigers are weaned by 6 months; domestic cats, at 7 weeks.

Cats' ability to live without their mother's support is related to when their permanent teeth, especially their canines, are in place. In smaller cats, dispersal begins

While learning to hunt, young cats often stalk and "kill" inappropriate things—waving blades of grass, for instance, or their mother's tail. This young cheetah has captured a real treasure: an elephant dung patty. By watching what their mother eats, youngsters soon learn to hunt more nutritious fare.

soon after the canines are fully erupted, as little as a few days to a few months later. This interval is much longer in big cats. Tigers and lions are able to kill small animals with their baby teeth, but until their canines are fully erupted, they are ineffectual at killing large prey, and it is possible that they need time to practice using those large canines on large prey, unlike small cats, which may continue to hunt the same small mice or birds throughout their lives.

HOW DO CATS LEARN TO HUNT?

The most important thing in a young cat's life is to learn to hunt. And to become a successful hunter takes a lot of training and practice. The development of hunting skills follows a similar pattern in all cats that have been studied, from domestics to cheetahs and tigers. In all species, mothers play a major role in the education of their young, acting as both facilitators and teachers. When young are a certain age (about 4 weeks old in domestic cats and as old as 2.5 to 4 months in cheetahs), a mother begins to bring them the prey she has killed, so that they can watch her eat it and eventually begin eating the meat themselves. Next, she brings them live prey, either intact or disabled, and vocalizes to call their attention to the animal. The young play with the prey and eventually kill it; the mother helps only if the prey is about to escape, in which case she will retrieve it, or if the kittens lose interest before killing it. Among big cats, whose large prey is difficult or impossible to bring live to the young, this stage is omitted, and young see their mother kill only later, when they begin following her while she hunts. At first they merely watch, but eventually the mother lets them take a crack at killing. Subsequently perfecting their hunting skills takes weeks or even years.

Young cats appear to learn a lot about how to be good predators by watching, imitating, and simply being with their mothers. Tim Caro (1994) carefully documented this in studies of domestic cats in the laboratory and cheetahs in the Serengeti. His domestic cat study compared how kittens' responses to live prey were influenced by the presence or absence of their mother when the kittens were between 4 and 12 weeks old—the time when domestic cats are typically learning to be predators. Kittens under 8 weeks of age interacted with and caught prey more often when their mother was present and were more likely to watch the prey when their mother was also watching it. By 12 weeks of age, kittens with their mothers had killed mice more than five times more often than those without their mothers and had killed significantly more mice overall. Caro noted that mothers vocalized to encourage their kittens to interact with prey and adapted their behavior to how well the kittens were doing. Initially, mothers killed most of the prey, giving the kittens a chance to watch and learn. Gradually, the mothers killed the prey less often, leaving it to the kittens to kill and eat. Finally, when tested again at 6 months of age, kittens that had learned to hunt with their mothers present were much better at biting prey on the nape of the neck, the way domestic cats typically and most efficiently kill.

A cheetah mother and her cubs begin traveling together when the cubs are 6 to 8 weeks old, which is also when cubs first begin eating solid food from their mother's kills. But soon, when cubs are 2.5 to 3.5 months old, the mother begins releasing some of her prey, such as newborn gazelles and hares, in front of her cubs before she kills it. The cubs chase it and knock it down for 5 to 15 minutes, until the mother finishes the job so they all can eat. Up until cubs are 4 months old, mothers release about 10 percent of the prey they catch but end up killing almost all of it. By between 4.5 and 6.5 months, the cubs are getting better at suffocating prey, which is how adult cheetahs kill gazelles, and the mother is releasing nearly a third of her catches and killing them herself less and less often. At this stage, cubs often eat prey alive while another family member holds it down.

The mother is sensitive to the abilities of her cubs, releasing 30 percent of newborn gazelles and only 4 percent of subadults and adults, and is less likely to maim newborn gazelles than hares, which otherwise escape rather easily from clumsy cubs. This continues until the cubs are 8 to 10 months old, when the mother begins releasing fewer prey to the young, preferring to kill them herself. At about 6 months, cubs begin to initiate hunts themselves but seldom succeed in killing; as a result, the mother's success may be reduced because the cubs alert prey to their presence. Cubs begin to approach adult skill levels only when they separate from the mother, at between 16 and 21 months of age. Surprisingly, young cheetahs are still not very good hunters at this point. They continue to chase after inappropriate prey, such as

These young cheetahs seem more curious about a gazelle fawn than intent on eating it. Mothers among cheetahs and other cat species often release live prey in front of their young, the first part of their hunting education. For the largest cats—lions and tigers, which hunt big dangerous prey—this is not an option. Lion and tiger cubs won't have this experience until they begin traveling with their mothers.

zebras and jackals. It takes until they are about 3.5 years to learn to judge the best distance from the prey at which to begin their approach, and they gradually improve their ability to choose vulnerable prey and to stalk most effectively.

Many anecdotal observations in other cats indicate how important mothers are in helping their young acquire hunting skills. Lion cubs follow their mothers at 5 to 7 months, participate in hunts at 11 months, kill gazelles at 15 to 16 months, and kill larger prey after they reach about 2 years of age. Serval mothers call their kittens to join them at kills.

Tigers cannot begin to hunt large prey effectively until their permanent canine teeth come in, when they are between 12 and 18 months old. Before this, they are still dependent on their mothers to provide meals, although they do catch the occasional small mammal or bird; they do not become truly proficient hunters until they are 2 years of age or older, improving with practice as they age and grow. Tigers don't achieve their maximum body weight and muscle mass until they are about 5.

As in other cats, play is another important component of tiger cubs' training as predators. From 3 or 4 months of age, cubs play a lot at being predators. They use one another as mock prey during bouts of rushing, chasing, back straddling, pawing, and biting. The behaviors observed in play are like those involved in hunting and killing and usually occur in the correct sequence. For instance, a cub initiates a play bout with its sibling by staring at it, stalking it, rushing it, chasing it, straddling its back, and then releasing or biting it. Some steps may be skipped—for instance, the cub may stare, rush, and stop, or it may rush, straddle, and stop—but the order remains the same. These sequences are also performed more quickly than they would be in a normal hunting sequence and are repeated many times, with very short breaks between them. This may provide practice for a variety of different prey and hunting conditions, letting the cubs improve their timing and coordinate their movements with those of prey. This type of play is called predatory play.

Tiger cubs also engage in aggressive play along with the predatory play. This too is very similar to unplayful aggressive behavior between adults and may quickly shift into aggression between cubs. Aggressive play includes biting, pawing, and wrestling but usually not threats, vocalizations, or submission displays. The movements are also slower and less forceful than those of true aggressive behavior—the idea is for the cubs to practice fighting but not hurt each other. Along with fighting practice, aggressive play may be a way for cubs to size one another up: a dominant cub will likely turn out to be a dominant adult, and these play interactions may influence how the cubs interact with one another as adults. Similar aggressive play is seen in other cats as well. Cubs and kittens also play with other objects. They stalk flowers waving in the breeze, for instance, or small bushes, or their mother's tail.

WHERE AND HOW MUCH DO CATS SLEEP?

When they are not out and about hunting or patrolling territories, cats generally hole up in some relatively safe place, such as a thicket of dense vegetation, a burrow, a crevice among rocks, and the like. This is where they rest and sleep and where females have babies. Lions may be seen sleeping with reckless abandon, sprawled out in the shade; we see our pet cats do this sometimes too. But generally cats seek secure places to let down their guard.

Most mammals exhibit two kinds of sleep: light sleep and deep sleep, which are differentiated by the brain wave patterns seen in an electroencephalogram (EEG). In light sleep, there is very little brain activity, and the animal is in quiet repose. During deep sleep, also called REM sleep for the rapid eye movements that occur

With no danger of predators and bellies full of food, lions may sleep or rest up to 19 hours a day. Being big and killing big prey enable a lion to eat a huge meal, then not hunt again for several days. Small cats sleep and rest less, requiring much more activity to catch many small meals each day. When prey is scarce, however, lions have to work long hours too.

in this state, intense brain activity is observed. During deep sleep, mammals, including cats and people, twitch, jerk, change positions, and so on, but are totally relaxed and more difficult to arouse than during light sleep, although once aroused they are more alert than when coming out of light sleep. In adult cats, light sleep, REM sleep, and wakefulness follow a regular cycle, with light sleep occurring as a transition from wakefulness to deep sleep and wakefulness directly following deep sleep.

A newborn domestic cat does not experience the light sleep stage, going directly from wakefulness to deep sleep and dividing its time about equally between the two for its first few days. Gradually it adds light sleep to the mix until, as an adult, it spends about 35 percent of its time awake, 50 percent in light sleep, and 15 percent in deep sleep.

Mammals that have safe sleeping sites and are predators generally spend more time in REM sleep than mammals that are prey and lack safe beds. As expected, lions, tigers, and domestic cats are known to have a large proportion of REM sleep; they also have a relatively large proportion of light and REM sleep combined.

Another correlation observed across mammals is that the larger ones generally sleep less and have less REM sleep than the smaller ones. However, cats seemingly violate this rule, because lions are the least active of all cats, whereas ocelots, for instance, hunt actively 12 to 14 hours a day. But, outside the laboratory, scientists usually cannot tell whether a cat is in light or REM sleep or just resting (not active), so a lion may not be sleeping all the time it isn't moving around, and an ocelot may use every spare hour to get some shut-eye. Moreover, cats vary their activity time depending on the availability of prey. Where prey is abundant, lions average 19 or 20 hours of rest a day; where prey is scarce, 14 hours of rest may be the norm.

Although all mammals sleep at least a little, and long-term sleep deprivation can lead to death, the function of sleep in its two different forms is unknown. Some ideas about its function include tissue repair, energy conservation, recharging the supply of neurotransmitters, and learning and the formation of memories. Support for the last-mentioned idea came recently from a study of young domestic cats during the critical period for visual development. Scientists found that kittens that slept for 6 hours following visual stimulation formed twice the number of brain connections than those that were kept awake and either left in the dark or were examined immediately after the visual stimulation. Cats kept awake and exposed to an additional 6 hours of stimulation also showed fewer changes in the brain than those allowed to sleep. This study also found that light sleep is responsible for the effect. Increasingly, scientists are coming to believe that light sleep is crucial to learning and the formation of memories, a role formerly attributed to REM sleep. If the brain builds new connections during light sleep, then REM may function as a testing phase, when the animal checks out how far the work has gone, wakes up briefly to make sure all is well in the world, then drifts back to light sleep again.

DO CATS DREAM?

People dream only during REM sleep, but the function and cause of those dreams are yet unknown. Dreams may simply be the result of the intense activity of the brain during this kind of sleep. Cats in REM sleep certainly behave as if they are dreaming: they twitch their paws, whiskers, and tail and even vocalize. We have no way of knowing, however, whether their dreams are similar to the often-bizarre filmlike experiences we remember when we wake up. If dreams are merely side effects of the host of physiological events that occur during REM sleep, it seems likely that cats do dream. However, if dreams have functions in people, such as promoting creative thought through linking ideas in bizarre ways that are normally

Do cats think? These lions appear to be planning a hunt, and the tactics these lions will use in their cooperative hunt suggest that the hallmarks of thinking—intention, planning, foresight, and flexibility—are present.

suppressed by rational thinking, as proposed by one scientist, it is hard to imagine that they serve that function in cats. If cats do dream, whether they remember their dreams upon waking, even fleetingly as we usually do, is unknowable at this stage of our ability to measure and compare how the brain works. Even if they do remember their dreams, they are unlikely to spend any time interpreting them.

HOW SMART ARE CATS?

The measurement of intelligence is fraught with difficulty, even when we are dealing only with people. In fact many scientists now object to the notion that there is a single IQ (intelligence quotient) that reflects much more than the ability to take standardized tests. Instead, they recognize multiple types of intelligence that include linguistic intelligence; logical-mathematical intelligence; spatial intelligence; kinesthetic intelligence; interpersonal intelligence; and naturalist intelligence. Students of animal behavior confronted a similar issue many years ago when they tried to test the intelligence of diverse species with the same tests. Cats, for instance, would quickly learn to escape puzzle boxes for a food reward but did worse than pigeons when asked to press a lever for a reward. This didn't mean that pigeons were smarter than cats, but only that cats, given their naturally complex predatory behavior, don't expect to find prey in this predictably easy way.

Another way that scientists have attempted to compare intelligence is by measuring the encephalization quotients (EQs) of diverse species. The EQ is a number reflecting the increase in brain size over and beyond what is explainable by an increase in body size. Thus, human beings have a very high EQ, with a proportional brain size far larger than that of other primates. In turn, primates have higher EQs than carnivores, and carnivores have higher EQs than rabbits. This fits with our intuitive notions about intelligence: monkeys are smarter than dogs and cats, which are smarter than rabbits. Canids have higher EQs than felids, which may be related to canids' tendency to live in groups, allowing them the development of more social intelligence than felids have, which are mostly solitary. Also, different parts of the brain may be relatively larger or smaller among different groups of animals, indicating different abilities. The olfactory lobes of the primate brains, for instance, are relatively small, whereas those of carnivores are relatively large, and those of dogs are larger than those of cats. Thus carnivores have better "smell smarts" than primates, and canids are better in this way than felids. Felids have relatively large cerebellums, the area of the brain that coordinates movement on a minute-to-minute basis, not surprising in light of cats' hunting and killing skills, which depend on precision attacks to vulnerable parts of their prey. Thus cats have kinesthetic intelligence that may exceed that of average people. Similarly, cats must have good spatial intelligence that enables them to find the boundaries of their large home areas and to know where prey and water are likely to be found, where others of their species are living, and the best routes to take—and when to take them—to find prey and avoid predators.

Asking whether cats think is related to asking how smart cats are. Thinking is defined with such concepts as intention, planning, foresight, and flexibility in the face of obstacles or change. No domestic cat owner will tell you that his or her cat has not demonstrated all these capacities, and each will have one or more favorite examples to show how smart his or her cat is. Wild cats appear to have these same capacities. Our favorite example from wild cats is that servals learn to take advantage of car headlights in their illumination of prey crossing the road. Lions learn the location of carcasses potentially available to scavenge by noting the circling of vultures. Cats are also capable of learning by observation. Young cats, for instance, learn what is appropriate prey by watching what their mothers eat (see *How Do Cats Learn to Hunt?*).

DO CATS HAVE ENEMIES?

People, parasites, viruses, and bacteria—all may be the enemies of cats. But day to day, the worst enemies of cats are other cats and other carnivores! In general, any carnivore will kill another one; the larger ones kill both adults and young of a smaller species, and smaller ones may kill the young of larger species. The cats of sub-Saharan Africa are a great example. One study in the Serengeti found that lions, leopards, and spotted hyenas account for 68 percent of the deaths of cheetah cubs. Also, leopards and hyenas kill lion cubs that have been left alone, accounting for about 8 percent of lion cub mortality in the Serengeti. Throughout lions' and cheetahs' ranges in sub-Saharan Africa, if lion numbers are high, cheetah numbers are low. Cheetahs may be forced to hunt in the middle of the day to avoid lions, which tend to hunt at night.

In North America, pumas and coyotes are responsible for from 12 to 62 percent of bobcat deaths. Gray wolves have recently been reported to kill pumas where the two species are meeting for the first time in a century, in the Greater Yellowstone ecosystem, as a result of the gray wolf's restoration and the puma's expansion in the American West. In Europe, Eurasian lynx and gray wolves have similarly uneasy relations. In Russia, gray wolves have been observed to kill a lynx, and lynx have been found in the stomach contents of wolves. In general, where wolves are abundant, lynx are not, and vice versa.

In Asian habitats, tigers are the nemeses of leopards. Where tigers are many, leopards are few, and where the two overlap in range, leopards often adjust their activity times and prey on smaller species to avoid encounters with tigers. But even the mighty tiger is sometimes bested by pack-hunting dholes—Asian wild dogs that weigh a mere 20 kilograms or so—although not without cost to the dholes. For instance, Sunquist and Sunquist (2002) describe how one tiger pursued by a pack of 22 dholes killed at least 12 of them before collapsing.

Although predation is assumed to be a source of mortality for small cats and the babies and young of larger cats, it is rarely directly observed. However, there have been observations of pumas, Eurasian lynx, and Iberian lynx killing domestic cats, caracals killing African wildcats (*Felis silvestris*), and jaguars killing ocelots. Servals cower at the sight of hyenas and run at the sight of dogs. Jungle cats (*Felis chaus*) may be rare in some places thanks to predation by leopards. From a cat's point of view, nonfelid carnivores are also enemies that they kill. Pumas kill coyotes, for instance; and tigers, leopards, Iberian lynx, Eurasian lynx, and pumas kill domestic dogs. In the Russian Far East, young brown bears are attracted to carrion, and those that find a kill belonging to a tiger are quickly dispatched by the tiger.

One difference in outcome between these intracarnivore killings and regular predation on ungulates, rodents, and birds is that the killer doesn't necessarily consume the victim, especially if an alternative food source is at hand. This suggests that some of these killings are related to competition over food rather than to the direct acquisition of food. In many examples, like that of the tigers and brown bears, one of the carnivores involved is scavenging and meets the other at a kill. Both bobcats and coyotes may attempt to steal from a puma's kill and, in turn, be killed for their efforts.

Potential noncarnivore predators include snakes—a jungle cat was found in the stomach contents of a python—as well as other reptiles and large birds of prey. For instance, in South Africa, crocodiles kill leopards that come to drink at the water's edge.

ARE CATS ALWAYS AFRAID OF DOGS?

It's clear from the previous question and answer that felids and canids can be enemies, but who's afraid of whom depends on the species and the situation. For instance, by force of numbers, pack-hunting wild dogs in Africa and dholes in Asia can kill lions and tigers, but not easily; conversely, lions and tigers are also known to kill wild dogs and dholes. As noted in *Do Cats Have Enemies?* many cats, including pumas, kill domestic dogs, but humans use dogs to hunt and chase pumas and bobcats into trees, where they can easily be shot. Domestic cats generally avoid or run from domestic dogs, but if kittens and puppies are raised together, they often form and maintain a companionable relationship as adults.

DO CATS GET SICK?

All mammals get sick. Mammals acquire internal and external parasites, catch infectious diseases, both viral and bacterial, and are attacked by various kinds of fungi. Cats are no exception, and many of the viruses and bacteria that infect them are the same as, or related to, ones that infect people. For instance, various wild cats are reported to suffer from bubonic plague, anthrax, tularemia, salmonella, and tuberculosis, among the bacterial infections, and rabies and encephalitis among the viral infections. Ticks, fleas, lice, and other ectoparasites plague cats, and the endoparasites that bother cats include an array of nematodes (round worms) and cestodes (tapeworms). Of 92 samples of feces from five species of cats living in the same area of Thailand, 88 of them, or 96 percent, showed evidence of endoparasites ranging from a single species to as many as nine different species. All told, a recent survey found that 15 viruses, 11 bacteria, 9 protozoa, and 1 fungus are known to infect wild felids; species had recovered from about half of these, as detected in the antibodies in their blood serum. Veterinarians have also identified 258 genetic diseases or abnormalities in domestic cats, including diabetes, hemophilia, retinal degeneration, and spina bifida.

Several diseases are unique to cats. Feline respiratory infections are caused by several different viruses or bacteria; the most common of these infections are feline viral rhinotracheitis, a herpesvirus, and calcivirus. Called cat "flus" for their symptoms, which include runny nose, sneezing, fever, and loss of appetite, they are not usually fatal in domestic cats. Another cat flu, also called feline pneumonitis, is caused by the bacterium *Chlamydia psittaci*. The primary symptom is conjunctivitis, or "pink eye." People may also be infected with *Chlamydia*.

Feline infectious peritonitis (FIP) is caused by a coronavirus infection. Many coronaviruses infect cats, but most do not cause serious disease. The mutant strain that causes FIP, however, invades white blood cells and spreads the virus through the cat's body. The immune response to the virus results in inflammation, and this interaction creates the disease. Interestingly, a cat's first exposures to FIP may result in mild symptoms or none at all. Weeks, months, even years later, the disease flares up in a small percentage of infected animals, and death is inevitable. Young, old, and stressed cats are the most susceptible to the disease. In domestic cats, animals that succumb are often also infected with feline leukemia virus. Lions, leopards, pumas, bobcats, lynx, and cheetahs are known to suffer from FIP, and cheetahs are particularly at risk.

Feline leukemia virus (FeLV) occurs in three forms. One weakens the immune system, another causes tumors and other abnormal growths, and the third causes severe anemia. FeLV kills more domestic cats each year than any other disease, but it has been reported in the wild only in European wildcats (*Felis silvestris silvestris*), which are in close contact and often interbreed with domestic cats. One study found that in domestic cats and wildcats, adult males were much more likely to carry both FeLV and feline immunodeficiency virus. Feline immunodeficiency virus (FIV) is very similar to HIV, the virus that causes AIDS in people, and FIV disease follows a similar course in its felid victims. In fact, FIV provides an animal model for studying these kinds of diseases. In addition to domestic cats, FIV is reported in lions, cheetahs, pumas, bobcats, leopards, and Pallas' cats, but each becomes infected with a species-specific strain of FIV.

Cats are the definitive host of the protozoan *Toxoplasma gondii*, which reproduces only in cats. Cats usually ingest this parasite with a meal, because a special form of the protozoan lives in meat. Released into the intestines, it reproduces and forms eggs, which are carried to the outside world in the feces. From the feces, other animals pick up the eggs, which resist drying and other environmental insults; they are also spread by wind and water. *T. gondii* can infest more than 200 other species of mammals and birds, taking up residence in various organs and creating symptoms that vary, depending on the infested organ, which may be the brain, liver, lungs, or other organ system.

Although *T. gondii* reproduces in cats' intestines, as in other species it can also infest nonintestinal organs. Domestic cats generally show only mild signs of infestation, but wild cats may become obviously sick, with Pallas' cats appearing to be extremely susceptible to toxoplasmosis, possibly because the protozoan was absent from the Pallas' cat's harsh habitat during the evolution of the species. An individual host may harbor dormant eggs for a long time until the immune system is compromised for some other reason, such as FIV, allowing the parasite to multiply. If a woman contracts *T. gondii* during pregnancy, the protozoan may cross to the fetus, which can be seriously affected because its immune system is not yet fully developed.

While the diseases of domestic cats and, to a lesser extent, some cats commonly kept in zoos are well documented, little is known about disease processes in cats in the wild. To avoid being easy targets for predators or competitors, wild animals hide signs of illness as best they can; cats, being already highly secretive animals, seldom provide us with observations of themselves when sick. Finding a dead cat in the forest still in a condition in which the cause of death can be determined is very rare. Scientists can infer that a cat has been exposed to a disease by measuring the presence of antibodies in the blood, but antibodies do not necessarily mean that the animal has been clinically ill; the presence of antibodies could mean instead that the animal fought off the disease.

In 1994, scientists had a rare chance to observe firsthand an epidemic of canine distemper virus (CDV) in lions in the Serengeti. Tourists first observed lions having convulsions and then alerted veterinarians, who rushed to the scene and eventually identified the illness as CDV. Usually a disease of domestic dogs, scientists believe that hyenas caught it from domestic dogs and passed it on to the lions. About one-third (1,000 of 3,000) of the lions died. The survivors, scientists hope, should be immune to the disease. To fight this threat, biologists are also trying to eliminate the reservoir of the disease in the local village dogs by vaccinating them. Although this epidemic exacted very high mortality, George Schaller estimated that typically only 18 percent of lion deaths in the Serengeti are due to disease.

Fortunately, the Serengeti lion population was very large and robust. However, conservation biologists are concerned that CDV will spread to small and especially vulnerable felid populations. For example, the 30 or so surviving Amur leopards (*Panthera pardus orientalis*) in the Russian Far East live in forest that is bordered by numerous villages, and leopards sometimes come into these villages and kill dogs to eat. The presence of CDV could threaten the survival of this entire leopard population.

While we intuit that rabies is a major threat to all carnivores, we have few records of rabid wild cats or confirmed deaths from rabies. However, the rabies antibody has been found in the blood of wild cats, so we know these individuals were challenged and recovered. It is unlikely that wild cats would infect one another because of their solitary life style, but they may be exposed to rabies in domestic dogs, other canids, or even other wildlife species that serve as reservoirs for the rabies virus, such as bats.

Cats' well-being can also be indirectly affected by disease. The most endangered cat in the world, the Iberian lynx, is critically endangered in large part because of the disease-caused decline of its main prey, the European rabbit (see *What Are the Most Endangered Cats?*).

DO CATS HAVE NINE LIVES?

According to Desmond Morris (1996), the nine-lives expression is based on two things: the domestic cat's amazing ability to escape from peril, and the tradition of a trinity of trinities in some religions, making nine a lucky number, which was then associated with the lucky cat. Of course, cats' success as escape artists is a result of skill, not luck. Alert senses, speed and agility, quiet movement, and camouflage coats enable cats to disappear in the face of danger when discretion is the better part of valor. When it is not, their powerful limbs, long claws, sharp canines, and swift attack capabilities enable cats to put up fierce a defense. Yet, despite their skills and capabilities, many wild cats enjoy barely one life.

In cats that have been studied, cub mortality if often quite high, with losses due to starvation, disease, predation, infanticide, and natural disasters such as floods and fires. And risks continue throughout life. Apart from the threats of starvation, disease, and predation, adults may be injured by prey and, especially among males, in fights over mates and territories. Human-induced mortality, from hunting, trapping, poaching, poisoning, and vehicles, is also high in some cat populations.

Despite their immense strength, lions do not have an easy life in the wild. They suffer from parasites and disease, they can be injured or even killed while hunting or fighting with one another, and they may starve when food is scarce. About two-thirds of all lion cubs die before they are 1 year old. Adult males are usually old and battered by age 10, if they survive that long, and they rarely live longer than 12 years. Females may live longer, up to 18 years, and some are still breeding at 15.

Tigers can live up to 15 years in the wild, but most do not. Only half of all tiger cubs survive to the age when they become independent of their mother. Only 40 percent of these survivors live to establish territories and begin to produce young. Among these territorial adults, the risk of death remains high. Males typically do not live as long as females, because they are more likely to engage in violent fights with other tigers to protect or gain territories. Adult tigers also suffer from parasites and disease, they may be injured or killed by a blow from a prey animal's flailing hoof, and they may starve when prey is scarce.

In Tanzania's Serengeti National Park, a cheetah cub has about a 5 percent chance of surviving to adolescence, which begins at 14 months of age. More than

70 percent of litters die during their first 8 weeks of life, mostly because of predation by lions, but also because of fire, exposure, and maternal abandonment. The next risky period occurs between leaving the mother and reaching adulthood, especially for males. Half of males die during this difficult time. In South Africa's Kruger National Park, about 50 percent of leopard cubs die in their first year, a third of these from predation and the rest from starvation.

DO CARS KILL MANY WILD CATS?

Cars kill many wild cats when they try to cross roads, even large cats such as pumas and tigers. Cats simply do not do well when they encounter cars traveling at high speed. They freeze in the headlight glare and are killed. In south Florida, the biggest source of mortality for the endangered Florida panthers (a subspecies of the puma, *Puma concolor coryi*) is cars hitting them on highways. The same fate awaits bobcats. Biologists attached radio transmitters to bobcats living near roads in south Florida. In every case, if a highway cut through the bobcat's home range, a car eventually killed the cat. The only bobcats to survive were those whose home ranges were not divided by highways. Iberian lynx have the same problem. Much of what we know about the food habits of South American cats has come from cats that were collected as roadkills. Tigers are reportedly being killed in India and the Russian Far East as roads are improved and traffic speeds up.

Roads are ecologically important to cats in another way: some cats are attracted to them, or rather to the prey that feed on the vegetation growing up in the disturbed conditions that road construction creates in the adjacent forest. This is a particular problem for tigers that feed on wild pigs and large deer. Not only are these ungulates more abundant near roadways, but also, tigers, being sight hunters, can scan long distances along road rights-of-way and thereby improve their hunting efficiency. Also improved, however, is the hunting efficiency of poachers. From far away, the headlights of a vehicle or a spotlight can aid poachers in detecting the eyeshine of a tiger walking along the edge of a roadway. They easily kill the cat with a high-powered rifle equipped with spotting scope. In the Russian Far East, all of the tigers that have been fitted with radio transmitters and live near roads have been killed in this manner.

HOW LONG CAN CATS LIVE?

Two different measures are used to determine a particular group's life span: one is maximum longevity, which is how long an individual in that group can potentially live; the other is life expectancy, which is how long an individual would normally live. In the United States, human maximum longevity is about 95 years, with the exceedingly rare individual, usually female, living to 110 or even 115 years. Average life expectancy, however, is about 77 years, a number that has shot up from 47 years at the beginning of the last century, thanks to advances in medicine. Too few data exist for any wild cat species to compute average life expectancy, but it is clear from *Do Cats Have Nine Lives? Do Cats Get Sick?* and *Do Cars Kill Many Wild Cats?* that few wild cats survive to adulthood, far fewer to old age.

Among domestic cats, the expected life span is between 9 and 15 years, as reported by Desmond Morris (1996), but a poorly documented record exists for a maximum longevity of 36 years and a more reliable record for 34 years. Feral domestic cats may have a much lower life expectancy. Among big cats, reported maximum longevities, mostly seen in zoo animals, are in the 20s—for instance, about 26 years for tigers, 22 for jaguars, more than 23 for leopards, and nearly 30 for lions. Records for other cats tend to fall in the high teens.

DO DOMESTIC CATS ALWAYS LAND ON THEIR FEET?

Domestic cats' reputation for having nine lives stems in part from their ability to survive long falls, and they do this remarkably well.

In 1987, veterinarians at the Animal Medical Center in Manhattan tracked the survival rates of 115 domestic cats whose owners brought them in for treatment after the cats had fallen from high-rise apartment buildings. Of the 115 cats, which had hit the pavement after falling from 2 to 32 floors, an astonishing 90 percent survived. The most intriguing part of the results was that cats that had fallen from 7 to 32 stories up died only half as often as cats that had fallen from 2 to 6 stories up. Several factors account for both these high survival rates in general and the even higher survival rates after falling the greater distances.

First, cats have an excellent sense of balance and exhibit what is called a righting reflex. Like a gyroscope, a cat's inner ear detects its orientation in space. As a cat falls, this gyroscope tells the cat to straighten up, and the cat does so by twisting its body in one direction and its tail in another, then stretching its legs and reversing the twist. The stretched out legs keep it from swinging back to the original rotation before its body and tail can twist in the opposite directions. The cat repeats this action until it is right side up, which it can usually achieve within the first 2 to 3 feet of a fall. Thus, cats land on their fronts, not on their backs.

Second, a cat is relatively small. This means that its bones, relative to its body size, are actually stronger than those of larger animals and are thus less likely to break under stress. Their small size and light weight also cause cats to fall more slowly than larger animals, so their maximum falling speed, known as terminal velocity, peaks at a lower speed than that of larger animals. A cat's terminal velocity is about 60 miles per hour; a person's is about 120 miles per hour. Therefore, a cat hits the ground more softly than a person does. Cats further reduce the impact by slowing their fall after reaching terminal velocity. To do this, they relax and spread their legs widely, creating a parachute-like effect comparable to the gliding of flying squirrels; this also spreads the impact of the landing over a larger body area. Finally, they land with legs flexed, not straight, which also reduces the force of the impact. This helps explain why longer falls may do less damage than shorter ones. Cats reach terminal velocity after about five floors, so if they land sooner they don't have time to break their fall by parachuting or to spread the impact throughout their body. By seven floors, they've relaxed, stretched out, and flexed their legs, giving themselves the best physical conformation to reduce the risk of death.

There is one flaw in the original Animal Medical Center study, and it relates to how the scientists measured mortality rates by looking only at the survival of cats whose owners sought treatment. What of the unknown number of cats that fell from high-rises and died instantly or at least before any veterinary help was obtained? Moreover, many of the cats in the study had serious injuries that may have been fatal without treatment. If these deaths could have been taken into account, the survival rate after falls would have dipped below 90 percent, but how far below is unknown. This flaw, however, probably does not change the conclusion that cats can survive longer falls better than shorter ones.

DO WILD CATS LIKE CATNIP?

That domestic cats go wild with a whiff of catnip is well known, and our awareness of this accounts for the sale of a lot of cat toys. Less well known is that catnip actually sends only about half of domestic cats into a frenzy of rolling, rubbing, and head shaking; the other half are indifferent to the herb and its active ingredient, a chemical called nepetalactone. The presence or absence of a dominant gene determines whether or not a domestic cat will respond to catnip. But why this herb induces a state that most closely resembles a female cat in estrus—although both male and female cats respond the same way—remains a mystery.

Scientists have tested how some other cats respond to catnip and have obtained mixed results. In captivity, tigers are indifferent to it, as pumas and bobcats are, even though hunters claim catnip oil bait attracts pumas, bobcats, and Canada lynx in the wild. Scientists trying to collect hairs for DNA analysis found that a commercial bait whose ingredients include catnip attracted pumas to rub stations, where they left behind a few hairs. However, they noted that unless the bait stations were refreshed regularly with novel odors, the pumas began to ignore them.

On the other hand, lions, snow leopards, and jaguars respond like domestic cats, and leopards show a weak attraction. As in domestic cats, males and females respond equally in the species that respond to catnip at all. People who keep servals as pets report that these cats enjoy catnip, but they have not described the cats' response.

Catnip, or catmint, is an herb native to Asia and Europe and is widespread in North America, where it was introduced long ago. But other plant chemicals are now known to produce a similar ecstatic response in domestic cats. For instance, a chemical called actinidine, from a kiwi relative called *Actinidia polygama*, works even better than catnip.

Offering novel scents to cats in zoos has become a part of programs designed to enrich the lives of these animals in otherwise fairly dull environments. Cats show interest in many spices, including mace, allspice, cumin, nutmeg, clove, and peppermint as well as musk in various perfumes, but not with the same intensity as some cats respond to catnip; their interest in these other scents wanes with familiarity, which seems not to occur with catnip. Cats' responses to perfume inspired scientists to use perfume, specifically Calvin Klein's Obsession for Men, to attract ocelots to camera traps. And it worked! The ocelots responded like cats in estrus. A female cheetah at the Bronx Zoo also found Obsession irresistible; for as long as 7 minutes, she would roll and rub against objects sprayed with it.

CLOCKWISE FROM TOP: Female cheetah and cub, *Acinonyx jubatus*; **jaguarundi,** *Puma yagouaroundi*; Iberian lynx, *Lynx pardinus* (PHOTO BY JOHN SEIDENSTICKER); **puma,** *Puma concolor*

CLOCKWISE FROM TOP LEFT: **Bobcat,** *Lynx rufus;* **Canada lynx,** *Lynx canadensis;*
clouded leopard, *Neofelis nebulosa;* **Eurasian lynx,** *Lynx lynx*

CLOCKWISE FROM TOP LEFT: Female lions, *Panthera leo;* **tiger,** *Panthera tigris;* **male lion,** *Panthera leo;*
snow leopard, *Uncia uncia*

CLOCKWISE FROM TOP LEFT: Leopard, *Panthera pardus*; **fishing cat,** *Prionailurus viverrinus;* **jaguar,** *Panthera onca;* **leopard cat,** *Prionailurus bengalensis*

CLOCKWISE FROM TOP LEFT: African golden cat, *Profelis aurata*; bay cat, *Catopuma badia*; caracal, *Caracal caracal*; flat-headed cat, *Prionailurus planiceps*

CLOCKWISE FROM TOP LEFT: **Marbled cat,** *Pardofelis marmorata;* **serval,** *Leptailurus serval;*
rusty-spotted cat, *Prionailurus rubiginosus;* **Asian golden cat,** *Catopuma temminckii*

CLOCKWISE FROM TOP LEFT: Sand cat, *Felis margarita*; jungle cat, *Felis chaus*; black-footed cat, *Felis nigripes*; domestic cat, *Felis catus*; wildcat, *Felis silvestris*

CLOCKWISE FROM TOP LEFT: Pallas' cat, *Felis manul*; **kodkod,** *Leopardus guigna*; ocelot, *Leopardus pardalis*

.2.

CAT EVOLUTION AND DIVERSITY

WHEN AND WHERE DID CATS EVOLVE?

Biologists try to understand why we have so many or so few species in one place or another. They ask how patterns in the history of Earth mesh with those in the history of a group of animals and how forces, conditions, and processes result in the diversity of species on Earth. In tracing the evolution of cats, or any animal, biologists are guided by the theory of common decent and evolution through natural selection. They look at the basic body plan and diagnostic characteristics of the cats alive today and trace them back to the first appearance in the fossil record of an animal with these characteristics; it is like starting with the tips of the branches of a tree and working down to the trunk. Periods during which little environmental change occurred in terms of climate, competitors, and predators produced fossils that exhibit little change as well. One kind of change occurs in response to other kinds of change. Over time, whether climates change slowly or abruptly, new habitats emerge as a result. Prey species evolve in their adaptations to exploit these new habitat types. The evolution of their predators is close behind.

The real Age of Mammals began after the extinction of the dinosaurs, about 65 million years ago, although mammals first emerged in the late Triassic more that 208 million years ago. The loss of the dinosaurs opened up ecospaces, or niches, for other creatures to proliferate. The major event triggering this shift was a huge meteorite that slammed into the Earth, probably just off the coast of Yucatán, Mexico. The order Carnivora arose from a miacoid, one of a group of small (1 to 3 kilograms)

Lions first appeared in the fossil record perhaps 3.5 million years ago in Tanzania, but the first fossils believed to represent a member of the family Felidae date to about 30 million years ago in France. Ancestors of most modern cats arose in the last 5 million years.

domestic-cat-sized arboreal carnivorans that first appeared in the forests of the Northern Hemisphere 60 to 80 million years ago. These early carnivorans had evolved a trait that was to give the group a flexible advantage in the drama of life that has been playing out ever since. They had evolved the key tool (and the diagnostic character) of the order Carnivora: a single pair of cutting teeth called carnassials, which are the upper fourth premolars and the lower first molars. With the enhancement of the cutting or the crushing aspect of these teeth, different groups of the Carnivora have become either meat-eating specialists, such as the cats and weasels, or plant-eating specialists, such as the red panda and giant panda, or a little of both, which describes most of the Carnivora species.

By about 40 million years ago, the first Carnivora species had divided into two sister groups. The catlike Feliformia eventually came to include the cats, hyenas, civets, mongooses, and a now-extinct catlike group called the nimravids. The second group was the bearlike Caniformia, which eventually came to include the bears, raccoons, weasels, dogs, skunks, badgers, sea lions, seals, walruses, and a now-extinct group called the bear dogs (Amphiconidae). By 37 million years ago, the civet-mongoose group, or clades, had diverged in their specializations from the cat-hyena group. By 35 million years ago, the cats and hyenas had separated into their respective clades.

The dawn cat, or *Proailurus*, was bobcat-sized, and its morphology suggests it was mostly arboreal. Fossils of this species are rare, but it or a loosely related group of species lived in Europe through the Oligocene (about 33 to 23 million years ago) and on into the mid-Miocene (about 17 million years ago). Cat fossils are very rare or absent in the early to mid-Miocene (about 23 to 17 million years ago), so biologists call this the cat gap. The cat ecospace, that of a stalking, pure meat-eater, was filled in North America by catlike forms of canids, bear dogs, and bears, which had teeth specialized for pure meat-eating. In Eurasia, canids, bear dogs, and hyenas evolved to use the cat ecospace. Fossils of *Pseudoaluria*, related to but distinct from the dawn cat, first appeared at widely separate sites in Europe, China, and North Africa during the early Miocene. They abruptly appear in fossil beds of the mid-Miocene in the Middle East, Mongolia, and India. This genus entered North America by 16 to 17 million years ago. There was a mid-Miocene decline in the catlike Nimravidae, the "old cats," or paleofelids. For the remainder of the Miocene, there was an increase in true cat fossils—members of the Felidae, the "new cats," or neofelids—and a decrease in the fossils of catlike forms from the other Carnivora families, presumably because true cats were better at being catlike and outcompeted the catlike forms that had evolved in the other Carnivora families.

During the Miocene, the Earth's climate changed, becoming drier and more seasonal. Shrub-grassland steppe, habitat like that of much of the intermontane West of the United States, and open grasslands, like those of the Great Plains or the

Serengeti, first appeared, as did grazing mammals adapted to these new habitats, greatly changing the nature of potential prey. A steadily increasing number of cat species evolved in response to these changes. A continuum of habitats from forest to forest-edge, to savanna, to open expanses of grassland and shrub created conditions in which different populations of a species evolved specializations for each different habitat, but there was still genetic interchange among populations. Prey adapted to these new conditions as well. Eventually, however, habitats became more distinct and diverse, resulting in the isolation of some peripheral populations of both predator and prey. With contact between populations broken, new species emerged.

The evolutionary history of successful carnivores is written in their attempts to overcome diverse prey animals, just as that of prey is written in their attempts to escape predation. In the forest, the herbivorous leaf-eating and fruit-eating species that are potential prey for cats do not have access to the majority of the plant biomass because it consists of tree trunks. With the evolution and expansion of grasslands and shrub, a whole new food source appeared for animals that were potential prey for cats. Ungulates and rodents emerged from the forest and evolved to eat the grasses and seeds of grasslands (see *Why Don't Cats Eat Fruits and Vegetables?*). Larger size and speed evolved among prey to enable them to escape predators in open environments. Predators evolved larger, faster body types in response. Among the cats, lions, cheetahs, pumas, sand cats, black-footed cats, and others hunt in open country, although they have greater hunting success where they can use some cover to stalk prey. Other cats, such as tigers and leopards, live at the forest edge but often hunt in adjacent open areas. Still others, such as margays and clouded leopards, are confined to forest.

Using a molecular genetic phylogenetic reconstruction—that is, analyzing genes to determine evolutionary history—Stephen O'Brien and his team at the National Cancer Institute's Laboratory of Genomic Diversity have traced the echoes of the divergence of the various groups, or lineages, of cats we know today back more than 10 million years to the late Miocene. This was a very dramatic period in the history of the carnivores. A major "turnover" occurred during which 60 to 70 percent of the Eurasian Carnivora genera and 70 to 80 percent of the North American ones became extinct and were replaced with large felids, canids, and bone-cracking hyenas. The last catlike nimravids vanished at this time. This upheaval is correlated with significant climate change that resulted in the desiccation of the Mediterranean Sea and the spread of seasonally arid grasslands in place of more moist woodlands. Most of the ancestors of the living cats first appeared in the Pliocene (5 to 1.8 million years ago) and the Pleistocene (1.8 million years ago to about 10,000 years ago). The relationships and the origins of the eight cat lineages are shown in Figure 4.

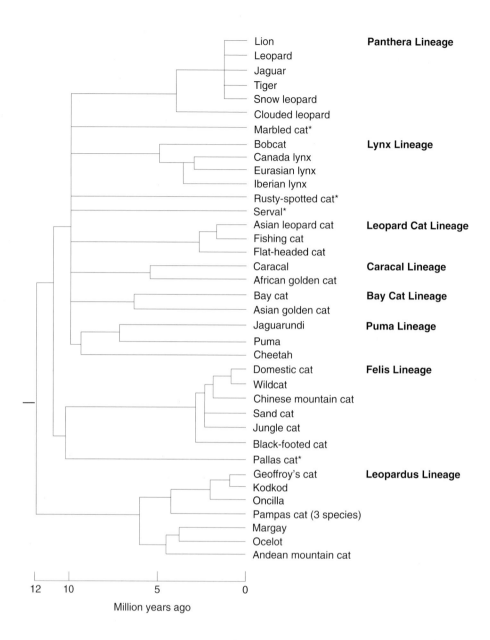

Figure 4. Several genetic analyses were combined to produce this tree, which reflects current understanding of the evolutionary relationships among the species in the family Felidae. The species within a lineage share a common ancestor, but in the Panthera Lineage, for example, clouded leopards diverged early from the ancestor of the other five cats in this lineage. Similarly, the tree shows that the Leopardus (Ocelot) Lineage diverged very early on from the rest of the cat lineages. Species marked with an asterisk did not cluster in any lineage consistently. (Iriomote cats were not included in the analysis that produced this tree, but they most likely fall into the Leopard Cat Lineage. In the absence of genetic analysis, the placement of the Iberian lynx within the Lynx Lineage was inferred on the basis of morphology.) (Adapted from Johnson, Dratch, et al. 2001)

HOW DO DIFFERENT CAT SPECIES EVOLVE?

The evolution of species within a region depends on the geographical configuration of the region, spatial arrangements of habitats, and the ability of the species to disperse into and out of that region. The formation of a new species, or the process of speciation, is the full sequence of events leading to the splitting of one population of organisms into two or more populations that become reproductively isolated from one another. New species can originate from geographically isolated portions of the parental species through their acquisition of an effective isolating mechanism. This is called allopatric, or geographical, speciation. There are two forms of allopatric speciation. The first form, called dichopatric speciation, involves the origin of a new species through the division of a parental species by geographical, vegetational, or other extrinsic barriers.

The flooding of the Bering Strait, between Asia and Alaska, at the end of the Pleistocene probably separated populations of a single species of lynx and created the Canadian and Eurasian lynxes. Advancing glaciers in the Pleistocene separated widespread populations into isolated ice-free refuges, and during arid periods of the same epoch, the large expanses of tropical rain forest were reduced to a number of smaller rain forest refuges surrounded by arid, treeless areas. We don't know how these may have influenced the number of cat species we see today, but we strongly suspect they did, especially in today's tropical areas in South America, Africa, and Southeast Asia. Retreating, melting glaciers resulted in rising sea levels, which created islands. This is probably what created the Bay Cat Lineage, with the bay cat currently restricted to the island of Borneo and its close relative, the Asian golden cat, living in much of the rest of tropical Southeast Asia.

The second form of allopatric speciation, called peripatric speciation, involves the origin of a new species through the modification of a peripherally isolated founder population. The founders may be just a few individuals or even single pregnant female. Some of these founder populations go extinct; some may eventually merge again with the parent species; some become full species. This is called species budding. A founder population may find itself in an entirely new environment or in a changing environment that offers the ideal situation for evolutionary departures into new niches and adaptive zones. The evolution of lions, cheetahs, and snow leopards may be examples, and so are the members of the Caracal Lineage. The African golden cat, a member of this lineage, is restricted to the rain forest, and the closely related caracal lives in the drier portions of Africa and south Asia, but not in deserts.

Large gaps occur between populations of many cat species, such as fishing cats, leopard cats, and rusty-spotted cats, perhaps because the species became extinct in parts of a once-continuous range. It is also possible for members of a species to establish founder populations far away from the original population after dispersing across unsuitable terrain such as water, mountains, or inhospitable habitat. This is called vicariance speciation. These widely dispersed populations may eventually become full species themselves.

Sympatric speciation, or the splitting of a single breeding population into more than one species without geographical isolation, is not known for any birds or mammals.

WHAT ARE SABER-TOOTHED CATS?

Growing up, we were transfixed by images of saber-toothed "tigers" that appeared in books and magazines, such as the wonderful C. R. Knight painting of a larger-than-lion-sized *Smilodon fatalis* lording over its domain. This great saber-toothed cat is known from many fossils of animals trapped in the La Brea tar pits in what is now Los Angeles, although the *Smilodon* genus was widespread in both North and South America and became extinct only about 10,000 years ago. The term *saber-toothed* comes from the shape of their upper canines, which resembled the long curved sabers brandished by mounted cavalry. *Tiger* referred to their large size but is misleading because it gives the impression that this species had an ancestral relationship to the tiger we know today, though it did not. Most saber-toothed cats were smaller than tigers. Paleontologists call tigers conical-toothed cats, and all cats living today are conical-toothed. The upper and lower canines are about equal in size, are shaped like cones, and are not serrated. When biting down, the upper canines fit behind the lower canines. Conical-toothed cats use their canines to bite into the throat or neck of their prey to kill them by suffocation, or by crushing, or by separating the vertebrae to damage the spinal column. Saber teeth come in two types: scimitar teeth and dirk teeth, both elongated upper canines. Dirk-toothed saber-toothed cats had extremely long, narrow upper canines with finely serrated edges. *Smilodon* was a dirk-toothed cat. Scimitar teeth are shorter, broader, and have coarse serrations.

The gape of a lion's wide-open mouth is about 65 degrees. That of *Smilodon* was about 100 degrees, and the gape reached 115 degrees in a scimitar-toothed species such as *Homotherium serum*, many fossils of which have been found in deposits in Texas and Eurasia. *Smilodon* was a powerful predator that probably ambushed prey. Because its saber teeth were relatively vulnerable to breakage, it probably stabbed and sliced its prey in places where there was little risk of hitting bone, such as the esophagus-jugular area of the neck, as paleontologist Larry Martin (1980) reasoned, or the stomach area, as other paleontologists have argued. The saber-toothed cat also must have been a careful feeder and avoided any contact with bone that might damage or fracture its highly specialized killing teeth. It probably left behind substantial portions of the carcasses of the prey it killed and, thus, provided food for a whole guild of scavengers. The scimitar-toothed cat *Homotherium* had longer front legs and probably pursued prey for longer distances than the squat and powerful dirk-toothed *Smilodon*.

The saber-tooth adaptation was found in cats of various sizes, and they have a very broad distribution. Fossil saber-toothed cats have been found in North and South America, through Eurasia, and into Africa. Moreover, this adaptation has not been restricted to cats; the saber-tooth specialization appeared independently at least four times in examples of convergent or parallel evolution. Many of the members of the extinct Carnivora family Nimravidae had long canines, as did large marsupial carnivores that once lived in South America. The marsupial saber-toothed carnivore *Thylacosmilus*, from South America, dates from the Pliocene, 5 to 1.8 million years ago. It had saberlike upper canines parallel to descending bony flanges on the lower jaw; it also had short, powerful limbs, suggesting it had a stalking-ambush hunting style. The Nimravidae were once thought to be the direct ancestors of the Felidae, but this is no longer considered to be the case. Yet, the Felidae and the Nimravidae both included saber-toothed species. The extinct creodonts—the first large carnivores to emerge after the demise of the dinosaurs 65 million years ago—also had species with saber teeth. The creodonts were a sister taxon to the miacoids, the group that gave rise to the Carnivora.

How are the saber-toothed cats related to the cats we know today? The molecular phylogeny was worked out by painstakingly harvesting ancient DNA from the fossil remains of the *Smilodon* from La Brea and comparing this with DNA from living species. This remarkable work was done by Dianne Janczewski in Stephen O'Brien's lab (Janczewski et al. 1992). The work confirmed that the most recent saber-toothed cats were clearly members of the modern family Felidae, not the Nimravidae. What remains unclear is just how the modern cats and the recent saber-toothed cats, as exemplified by *Smilodon*, are related to one another. The saber-toothed cats have traditionally been included in their own subfamily, the Machairodontinae, within the family Felidae, and the conical-toothed cats in their own subfamilies, the Felinae and Pantherinae. These biologists propose instead that the old subfamily designations be dropped because all the cats are more closely related than we once thought.

For the last 40 million years or so, there has always been a saber-toothed mammalian carnivore living some place on Earth. Through much of their history, humans lived cheek by jowl with great saber-toothed cats, until the last saber-toothed cats, *Smilodon*, became extinct. We are now in what paleontologist Blaire Van Valkenburgh (1991) calls a unique saber-toothless period. Clouded leopards have

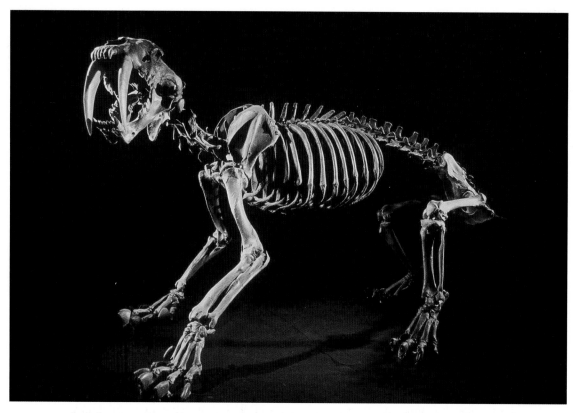

Smilodon was just one of many cats with saber teeth that once lived in North and South America, Eurasia, and Africa before the last of these species disappeared at the end of the Pleistocene. *Smilodon* has recently been shown to belong to the modern family Felidae. (Photo by Chip Clark)

very large canines for their body size, but these are of the conical-tooth type and thus different from *Smilodon* and other extinct saber-toothed cats.

WERE BIG CATS ONCE BIGGER?

The largest cat we have ever seen was a wild male Bengal tiger in Nepal. Male 105 he was called, because he was the fifth tiger captured in the Smithsonian-Nepal Tiger Ecology Project in 1975. He was massive. His muscles rippled, and he did not seem to have an ounce of fat. He weighed more than 250 kilograms; how much more we don't know, because the scales we had registered no higher than that. Reports exist of tigers weighing more than 300 kilograms, but it is likely that these weights were estimated and perhaps exaggerated. Other reports of heavier tigers have involved some very fat zoo animals.

However, some of the largest cat species in existence today were once larger, and other very large cats are now extinct. The fossil remains of American lions that were trapped in the La Brea tar pits more that 10,000 years ago show that those animals were huge. Some were estimated to have weighed 344 to 523 kilograms, more than twice the mass of a modern African lion (180 kilograms) or a very large tiger (250 kilograms). The great saber-toothed cat *Smilodon*, whose fossil remains were trapped in the tar along with the American lion, is estimated to have weighed between 347 and 442 kilograms. Another saber-toothed cat, *Homotherium serum*, is estimated to have weighed between 146 and 231 kilograms, a close match for today's African lion or Asian tiger.

Why were the big cats bigger in the past? Paleontologists speculate that their size allowed these very big cats to kill very large prey, such as giant bison, horses, mylodont sloths, camels, mastodons, mammoths, and tapirs, the fossils of which were also found in the same tar pits with their giant predators. This whole assemblage of giant mammals, predator and prey, is now extinct. Scientists argue heatedly about whether they were killed off by the first wave of humans into North America, suffered because the climate became drier and harsher, or some of both.

DID LIONS AND TIGERS ONCE LIVE IN NORTH AMERICA?

Lions once did range in North America, and they may have lived in South America too. In Alaska, fossils have been found of what may be tigers, but some biologists believe these fossils come from lions. Paleontologists are challenged by the difficulty of telling the big cats apart without their coats. Some biologists say it is impossible to tell the fossilized skulls of lions and tigers apart or even to distinguish them from leopards and jaguars. Size offers some clues: the largest leopard skulls today are considerably smaller than the smallest lion and tiger skulls, but this doesn't help separate lions and tigers. Lions, however, have longer legs than tigers relative to their body size, and where more complete fossilized skeletal material exists, the ratio of limb-bone length to skull size helps to separate the two species.

IN WHAT PARTS OF THE WORLD DO CATS LIVE?

Cats occupy terrestrial habitats virtually the world over. Only Australia, Antarctica, and islands that have never had a land bridge to a continent with a native cat species lack native wild cats. Because humans have left domestic cats wherever they have gone, feral domestic cats now live nearly everywhere. The general distribution and diversity of cat species in selected areas is shown in Figure 5.

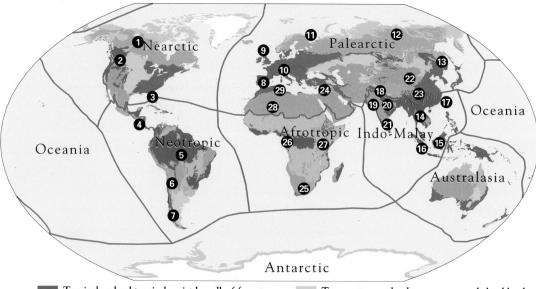

Tropical and subtropical moist broadleaf forests
Tropical and subtropical dry broadleaf forests
Tropical and subtropical coniferous forests
Temperate broadleaf and mixed forests
Temperate coniferous forests
Boreal forests / Taiga
Tropical and subtropical grasslands, savannas, and shrublands

Temperate grasslands, savannas, and shrublands
Flooded grasslands and savannas
Montane grasslands and shrublands
Tundra
Mediterranean forests, woodlands, and scrub
Deserts and xeric shrublands
Mangroves

NORTH AMERICA
1. *Boreal forest* / Canada lynx
2. *Idaho* / puma, bobcat, Canada lynx
3. *Florida* / puma, bobcat

CENTRAL AMERICA
4. *Costa Rica (not Panama)* / jaguar, puma, jaguarundi, ocelot, margay

SOUTH AMERICA
5. *Central Amazon rain forest* / jaguar, puma, jaguarundi, ocelot, margay, oncilla
6. *High central Andes* / Andean mountain cat
7. *Southern end* / puma, Geoffroy's cat, kodkod, Argentinean pampas cat

EUROPE
8. *Iberian Peninsula* / Iberian lynx, wildcat
9. *Northern Scotland* / wildcat
10. *Switzerland* / wildcat, Eurasian lynx
11. *Boreal forest* / Eurasian lynx

NORTHERN ASIA
12. *Boreal forest* / Eurasian lynx
13. *Russian Far East (temperate forest)* / tiger, leopard, leopard cat, Eurasian lynx

TROPICAL ASIA
14. *Thailand* / tiger, leopard, clouded leopard, Asian golden cat, marbled cat, leopard cat, fishing cat, jungle cat, flat-headed cat

15. *Borneo* / bay cat, clouded leopard, marbled cat, leopard cat, flat-headed cat
16. *Sumatra* / tiger, clouded leopard, marbled cat, leopard cat, Asian golden cat, fishing cat, flat-headed cat
17. *Iriomote Island (just east of Taiwan)* / Iriomote cat
18. *Nepal, lowlands* / tiger, leopard, leopard cat, fishing cat
19. *India, Gujarat* / lion, leopard, cheetah-E, caracal, wildcat, rusty-spotted cat
20. *India, Central M.P.* / tiger, lion-E, cheetah-E, leopard, jungle cat, leopard cat, rusty-spotted cat
21. *Sri Lanka* / leopard, fishing cat, rusty-spotted cat, jungle cat

CENTRAL EAST ASIA
22. *Mongolia* / snow leopard, Pallas' cat, Eurasian lynx
23. *Mountains of south-central China* / tiger-E, leopard, Asian golden cat, leopard cat, clouded leopard, Chinese mountain cat, jungle cat, Pallas' cat, snow leopard, Eurasian lynx

ASIA MINOR
24. *Israel* / lion-E, cheetah-E, leopard, wildcat, jungle cat, sand cat, caracal

AFRICA
25. *South Africa* / lion, leopard, cheetah, caracal, serval, wildcat, black-footed cat
26. *Rain forest, central* / African golden cat, leopard
27. *East Africa* / lion, leopard, cheetah, caracal, serval, wildcat
28. *Sahara, central* / sand cat, cheetah
29. *North edge, Africa* / lion-E, leopard-E, serval-E, caracal, wildcat

Figure 5. The comparative distribution of wild cats. In mapping the Earth's biodiversity, ecologists have described 14 biomes and 8 biogeographic realms. The cat species that now live at selected sites within these realms and biomes (or have been only recently extirpated in them, designated with E) are shown. (Biome map from the World Wildlife Fund–US)

The center of felid diversity and abundance is Asia, home to 22 species, with 10 species living in Southeast Asia alone and 10 species living in the mountains of China on the eastern side of the Tibetan plateau. Africa has 9 species. South America has 12 species. Today Europe has only 3 species (Iberian and Eurasian lynxes and wildcats), but lions lived in southern Europe in historical times. North America north of Mexico now has 4 species (puma, bobcat, Canada lynx, and a small population of ocelots in Texas). Within historical times, the jaguar and jaguarundi were also part of the North American assemblage, but today they are only occasional visitors.

HOW WIDELY DISTRIBUTED ARE VARIOUS CATS?

A few cat species have extremely wide geographical distributions. Leopards inhabit most of Asia and Africa, and pumas much of North, Central, and South America. The Eurasian lynx lives across northern Europe and Asia. Wildcats live in Europe, Africa, and Asia. The leopard cat's range stretches from the Indonesian islands through Southeast Asia, into part of India in the south, and into the Russian Far East in the north.

Other species have very restricted ranges. Bay cats live only on the island of Borneo; black-footed cats live only at the southern end of Africa; kodkod live in the southern Andes; and the Andean mountain cat lives only in the central Andes. Iberian lynx live only on the Iberian Peninsula; the African golden cat is restricted to African rain forest areas.

The distributions of the cats that are habitat specialists are easily read from maps of vegetation types and landforms. The African golden cat lives in Africa's wet tropical forests; its closest relative, the caracal, is found in the remaining areas of Africa that are neither wet tropical forest nor true desert, and it occupies similar habitats into Asia Minor and south Asia. The sand cat lives in the sand deserts of Africa and Asia Minor to Pakistan. The snow leopard lives in the mountains of central Asia. The Pallas' cat lives in the parts of central Asia that are dry and usually cold, but it does not live where there is any appreciable snow accumulation in winter. The Asian golden cat, marbled cat, and clouded leopard are closely associated with rain forest but also with the moist forests in the northern and western portions of their range in Asia. The bay cat and flat-headed cat are cats of the Southeast Asian rain forest. The oncilla and margay are tropical wet-forest cats of South and Central America. Canada lynx live in the boreal (northern) forest of North America and are tied to the distribution of their primary prey, snowshoe hares. The Eurasian lynx occurs across Eurasia, where it inhabits the boreal forest but also the margins of the steppe and the mountains. Hares are a major part of its diet, but it also feeds on small ungulates.

Some species that are widespread and even some with smaller geographical distributions have disjointed ranges, or vicariant distributions, that are the manifestations of past climatic and geologic changes. Today, the serval is confined to sub-Saharan Africa, but a North African population, separated by the Sahara, went extinct. The sand cat's range has been divided into at least five fragments of sand desert stretching from the western Sahara to Pakistan. We don't know why the populations have become fragmented. The leopard cat has a wide distribution, as we have seen. But where its range extends west along the foothills of the Himalayas, it suddenly stops and mysteriously does not occur through most of the rest of India except in the southwest, on the Malabar Coast. Fishing cats also have a disrupted distribution. They occur on Java and Sumatra but not on the Malay Peninsula or Borneo, then appear again in Southeast Asia and extend along the foothills of the Himalayas. They, too, are mysteriously absent through much of India except in the southwest, on the Malabar Coast. They are also found on the island of Sri Lanka. The westernmost population once occurred along the Indus River in Pakistan but is now believed to be extinct there.

Some species are undergoing massive range collapse. Lions now live only in sub-Saharan Africa, except for a small relict population in India's Gir Forest, but 12,000 years ago they were found at virtually every corner of the Earth, occupying the Americas, all of Europe, the Middle East, northern Asia to Siberia, and southern Asia to India and Sri Lanka. Cheetahs are also now found mostly south of the Sahara in Africa except for relict populations of a few tens of individuals in the central Sahara and in Iran. Formerly they had nearly a continuous distribution in open habitats throughout Africa, Asia Minor, the Middle East, and southern Asia, including the Indian subcontinent. The caracal's distribution once nearly mirrored that of the cheetah but is now limited to specialized habitat within its former range. In historical times, tigers lived across 70 degrees of latitude and 100 degrees of longitude, from the Russian Far East to the south through Indochina, the Indian subcontinent, and into the Indus Valley. Apart from this primary distribution, tigers lived adjacent to the Caspian Sea and on the Indonesia islands of Sumatra, Java, and Bali. Now they are confined to about 150 isolated patches scattered over this region, and those that lived around the Caspian Sea and on the islands of Bali and Java are extinct. On the other hand, the puma, which had been extirpated from all but the most remote mountain systems in the western United States, today is expanding into mountain ranges where it has not been seen for 100 years or more.

WHAT ARE THE RAREST CATS?

Conservation biologist Michael Soulé has pointed out, "Diversity and rarity are synonyms for 'everything' in ecology" (1986, 117). Ecologists who can explain and

In 1992, scientists saw a live bay cat for the first time ever; previously it was known only from 12 museum specimens. Since then, two have been photographed by camera traps. The bay cat, found only in the rain forest on the island of Borneo, is the rarest of the living cats. "Bay" refers to this cat's chestnut-colored fur, not its habitat.

predict patterns of diversity and rarity in landscapes or regions understand one of the most fundamental issues in biology. Biologists recognize different types of rarity. An animal may be called rare because it occurs only in one particular small area, or because it is restricted to one of a few specialized habitats, or because it is always found in small numbers across its range.

The rarest cat in the world is probably the bay cat, which lives both at a very low density and only on the island of Borneo, where it is apparently restricted to rain forest. The bay cat is known from only a few museum specimens and a very few individuals that have been captured and photographed. A camera-trap photo of a bay cat was taken in 2002 and another one in 2003. No individuals are living in zoos. Biologists believe this cat has always been very rare; they were not encountered during night forays into the field with spotlights except in one area of Sarawak (part of Borneo).

Just across the way, on the island of Sumatra, biologists regularly see and camera-trap the bay cat's close relative, the Asian golden cat. John has seen many tracks of Asian golden cats while walking along logging roads in the Sumatran rain forest.

The Iriomote cat is restricted to the small Japanese island of Iriomote, near Taiwan, where fewer than 100 occur. It is very closely related to the leopard cat, which is much more common and widely dispersed in Asia. The Iriomote cat eats almost anything it can catch and is not a strict habitat specialist.

The tiny rusty-spotted cat is called rare and has a patchy geographic range that extends from Sri Lanka to Kashmir, where there is only a single old record of its occurrence. Biologists didn't know it lived in the tropical dry forest of western India until A. J. T. Johnsingh photographed one in 1989. Subsequent searches resulted in more photographs of this little cat in and outside the Gir Forest in southern Gujarat. In southern India, villagers showed biologists where rusty-spotted cats were living in abandoned houses in thickly populated areas, distant from any forest thought to be their habitat. This cat may be more common than we appreciate.

The Andean mountain cat is also a very rare cat, known only from 14 museum specimens and three published observations as well as from some recent photographs of a few individuals. This cat is thought to have been much more abundant until a primary prey species, the chinchilla, was decimated by the fur trade.

WHAT CHARACTERIZES THE LINEAGES OF CATS?

Closely related groups of cats are called lineages, or clades, as shown in Figure 4. Relationships are determined by looking at molecular similarities of the genes and chromosomes of the various species. Using these methods, biologists have been able to resolve the cat lineages back to about 10 million years ago. Ancestors of some of the cats diverged before this time, and it is not certain how they fit into the currently recognized lineages. Despite this uncertainty, these cats are included with some of the lineages we describe below rather than being treated separately. Figure 4 traces the ancestors of today's cat species, but we cannot assume that the cats we know today are the same as their ancestors. Also, probably many other budding species arose but did not survive. There are many fossils of long-extinct cats, but most speciation events will never be known. All the cats we know today are the result of one long evolutionary process. In the formal classification of the living cats, the Panthera Lineage is included in the subfamily Pantherinae, and all the other lineages are in the subfamily Felinae, but some specialists think these distinctions are not correct.

Panthera Lineage: Clouded Leopard, Lion, Tiger, Jaguar, Leopard, and Snow Leopard, plus Marbled Cat

Often thought of as the archetypical big cats, the members of the Panthera Lineage vary markedly in weight. Lions (*Panthera leo*) and tigers (*P. tigris*) are extremely large, weighing more than 80 kilograms and up to 200 kilograms for the largest lions today and 250 to 300 kilograms for the largest tigers. A few of the very largest jaguars (*P. onca*) weigh as much as 100 kilograms. However, most jaguars, as well as leopards (*P. pardus*) and snow leopards (*Uncia uncia*), are medium-sized cats, weighing between 20 and 80 kilograms. The largest snow leopards are 50 kilograms; the largest leopards are 70 kilograms. While all the large cats are terrestrial hunters, the clouded leopard (*Neofelis nebulosa*), the smallest big cat, is also adapted for hunting in trees. Biologists say that the clouded leopard has a small cat body and a large cat head, with the longest canines of any cat relative to body size. Coat colors in this lineage match habitat, as they do in all the cats. Living in dry, open country, lions are tawny in color with a striking furry, dark tail tip. The forest-living tiger has black stripes on a reddish-orange background. The jaguar and leopard, also forest-living, have spots and rosettes; the jaguar has a central spot in its rosettes, and the leopard does not. The alpine snow leopard has longer, lightly spotted, smoky-gray fur with countershading. The forest-living clouded leopard has a blotched, countershaded coat.

The marbled cat (*Pardofelis marmorata*) is found only in the rain forests of Southeast Asia and diverged from all the other cats more than 10 million years ago. Molecular biologists are uncertain of its affiliation with any of the cat lineages, but some features suggest that it may be most closely associated with the Panthera Lineage, and with the tiger in particular. Marbled cats, thought to be arboreal, have never been studied, but they have been camera-trapped on the ground. They are extremely small, weighing from 2 to 5 kilograms. The marbled cat has thick, soft, spotted fur that varies from dark gray-brown, to yellowish-gray, to red-brown. These cats have large feet for their size and long bushy tails.

The clouded leopard's ancestor diverged from the rest of the lineage about 5 million years ago, at the beginning of the Pliocene. Molecular biologists tell us that the rest of the cats in this lineage diverged from one another in the Pleistocene, 1.8 million years ago, and more recently. However, paleontologists date some of the fossils back to the mid-Pliocene, about 3 million years ago or so, but because it is very difficult to tell species apart on the basis of skull characteristics, it's hard to say whether a given fossil is a very large leopard or a small tiger or whether it is a very large jaguar or a small lion (see *Did Lions and Tigers Once Live in North America?*).

Tigers live in forest and lush tall grassland and reach their greatest density at the edges between forest and grassland. The earliest fossils thought to be tigers appeared almost simultaneously about 2 million years ago in northern China and Java. However, it's not clear whether these were very small tigers, smaller than any we know today, or leopards that are larger than any we know today. By the mid-Pleistocene, the tiger had a wide Asian distribution, including China, Java, and Sumatra. The first appearances of the tiger in the Russian Far East and India occurred later. Fossil tigers exist from Japan in the late Pleistocene. Traditionally the tigers were divided into eight subspecies: Javan (*Panthera tigris sondaica*), Bali (*P. t. balica*), Sumatran (*P. t. sumatrae*), Indo-Chinese (*P. t. corbetti*), South China (*P. t. amoyensis*), Amur (Siberian) (*P. t. altaica*), Bengal (*P. t. tigris*), and Caspian (*P. t. virgata*). The Javan, Bali, and Caspian are extinct. The most recent genetic analysis indicates that the Amur tiger, the Sumatran tiger, and the tiger of the Malay Peninsula (formerly included as part of the Indo-Chinese subspecies) are actually distinct enough to deserve subspecies categorization (at least), and the remaining Asian tigers are lumped into a fourth subspecies.

Lions live in open plains or relatively open woodlands. They originally had the largest geographical distribution of any cat. The first fossils thought to be those of lions are from Tanzania 3.5 million years ago. Once the lion adapted to living in open country, it dispersed from Africa into Eurasia and crossed the Bering land bridge into North America, where it became widespread. However, the American lion disappeared about 10,000 years ago. Fossils thought to be those of lions have been found as far south as Peru, but some paleontologists now consider those fossils and others from South America to be very large jaguars. Except for one relict population of about 300 individuals that live in the Gir Forest of India, the once great geographical range of lions has shrunk to sub-Saharan Africa. Even within their present range, lions live in highly fragmented populations and are frequently killed outside protected areas.

Brian Kurten and Elaine Anderson (1980) have suggested that the New World jaguar is the same species as one known from fossils found in Eurasia of a cat that is now extinct. They suggest it crossed into North America about 800,000 years ago. These early jaguars were about 20 percent larger than the largest ones living today. There are jaguar fossils from many sites in North America from the Pleistocene, 1.8 million to about 10,000 years ago, when, like the lion and the saber-toothed cats, the North American jaguar went extinct. Today's jaguars recolonized Central and North America comparatively recently from refugia in northern South America.

The earliest fossils thought to be those of leopards are from Africa in the mid-Pliocene, 3.5 million years ago. Subsequently, the leopard dispersed through Eurasia. As many as 27 subspecies of leopards have been recognized, but genetic analysis suggests that only 9 are valid, including African, Javan, Sri Lankan, Amur (Siberian), India, Asia Minor, south Asia, Southeast Asia, and east Asia south of Beijing. The first snow leopard fossils are from Pakistan from more than 1 million years ago.

Both male and female lions live in social groups composed of close relatives, an adaptation for living in open country, where other lions are abundant and dangerous competitors. Lions kill medium-sized and large open-country ungulates, such as wildebeests and zebras. The other members of this lineage live in solitary social systems. The snow leopard is a specialized predator of goats and sheep through the rugged mountain ranges of central Asia. Leopards live in habitat as diverse as the Kalahari Desert, dry and wet forests, savannas, and woodlands, where they kill medium-sized ungulates and primates, supplemented with smaller mammals and birds. The clouded leopard is an enigma to biologists, because in captivity, unless young cats are reared together from before the age of 6 months, a male nearly always kills a female when they are placed together to mate. They prey on terrestrial species such as muntjac and small deer as well as tree-living monkeys and apes. Tigers are specialized predators of large deer and pigs and live in mangrove forests, rain forests, tropical wet savannas, and cold temperate forests. The jaguar lives in moist savannas and rain forest and kills deer, peccaries, and also caiman and large terrestrial and aquatic turtles.

Lynx Lineage: Bobcat, Iberian Lynx, Canada Lynx, and Eurasian Lynx

The four species of lynx are the foremost predators of hares and rabbits. On the basis of fossils, we now believe the Issoire lynx, more robust than today's lynxes, originated 4 million years ago in Africa, during the early Pliocene, and is thought to be the original lynx. The Pliocene (5 to 1.8 million years ago) was a period of glacial expansion and retreat, with a colder and drier seasonal climate than in previous periods, perhaps selecting for a lynxlike cat. But today no lynxes remain in Africa; this lineage is found only in Eurasia and North America. Pleistocene lynx fossils have come from many sites in Europe, Asia, and North America. Molecular genetic analysis shows that the North American bobcat (*Lynx rufus*) is the oldest branch of the radiation, followed in age by the Iberian lynx (*L. pardinus*). The Canada and Eurasian lynxes (*L. canadensis* and *L. lynx*) have been separated for about 3 million years, when rising sea levels submerged the Bering land bridge between Asia and North America. Naturally occurring hybrids of Canada lynx and bobcats have been reported.

All the lynxes are spotted and have soft hair, face ruffs, and black tassels on their ears. The Iberian lynx, Canada lynx, and bobcat are small (5 to 20 kilograms); most of the Eurasian lynx are medium-sized cats, weighing up to 40 kilograms. Lynxes are solitary, terrestrial hunters mostly specializing in rabbits and hares, so their evolution should correspond to the evolution of the lagomorphs, but biologists remain uncertain about the evolutionary history of lagomorphs. Rabbits and hares are thought to have last had a common ancestor about 15 million years ago. Today's hare species, such as snowshoe hares, jackrabbits, and European hares, are all placed in a single genus, *Lepus*, and can be traced back to the late Pleistocene. Some rabbits may have more ancient roots. In North America, the typical rabbit belongs to the cottontail genus, *Sylvilagus*; in Europe, there is one species—the European rabbit. (A third group of lagomorphs, the small pikas, aren't important in the lives of lynxes but are for the Pallas' cat.)

The Iberian lynx is a specialized predator of the European rabbit, which was originally from the Mediterranean region but has since been widely introduced elsewhere as well as domesticated by man. The Iberian lynx must have adequate populations of European rabbits to live (see *What Are the Most Endangered Cats?*). Cottontails and bobcats are closely matched in their distributions and habitats, although cottontails occur in the northern Neotropics, and bobcats do not. Where rabbits are few, bobcats live on a diet of rodents and birds. They are also significant predators of white-tailed deer in some areas.

Across the boreal forests of North America and Eurasia, the Canada lynx and Eurasian lynx hunt hares, and Canada lynx populations follow the cyclical waxing and waning of hare numbers (see *How Do Cat Numbers Affect Prey Numbers and Vice Versa?*). These northern lynxes are forest hunters and don't hunt hares that live in the open tundra. In Newfoundland and some other areas, Canada lynx eat caribou calves on the calving grounds and grouse and red squirrels when they find them, but their fortune is tied to the snowshoe hare. Both hares and lynxes have large feet, which are adaptations for moving easily through snow. In addition to hares, the larger Eurasian lynx hunts roe deer, red deer, and even wild pigs. In some areas, these ungulates are the mainstay of the lynx diet year around.

Leopard Cat Lineage: Leopard Cat, Iriomote Cat, Fishing Cat, and Flat-headed Cat, plus the Rusty-spotted Cat

In the Pliocene in Southeast Asia, less than 5 million years ago, the flat-headed cat (*Prionailurus planiceps*) diverged from the ancestor of the leopard cat (*P. bengalensis*) and the fishing cat (*P. viverrinus*). Then the ancestors of the fishing cat and the leopard cat diverged, and most recently the leopard cat and the Iriomote cat (*P. iriomotensis*) diverged. The Iriomote cat lives only on the island of Iriomote (see *What Are the Rarest Cats?*). The rusty-spotted cat (*P. rubiginosus*) diverged more than 10 million years ago, and molecular biologists are not certain how it is related to this or any of the other lineages, although biologists who base their work on morphology place it in the same genus as the others of the Leopard Cat Lineage.

The cats of this lineage are all spotted, with uniquely and beautifully marked faces. Their background coat is dark to light gray. The leopard cat and Iriomote cat have dense spotting on their coats; the others are less densely spotted. The flat-headed cat lives in tropical rain forest and also seems to do well where the rain forest has been converted to oil palms. The fishing cat is a cat of Southeast Asia, with separate populations in south India and Sri Lanka. It lives near water in rain forest and in moist forests. Recently John and his associates camera-trapped fishing cats living in Sri Lankan cities (Seidensticker 2003). The leopard cat is widespread in vegetation ranging from rain forest to dry thickets and woodlands in south and Southeast Asia, all the way into the temperate forests of the Russian Far East. The fishing cat, Iriomote cat, flat-headed cat, and leopard cat are primarily terrestrial hunters; the leopard cat, however, also can be a semi-arboreal hunter in some areas. The rusty-spotted cat, flat-headed cat, and the leopard cat living on the island of Borneo are extremely small, less than 3 kilograms in weight. Leopard cats in the

Russian Far East may be as large as 10 kilograms but in most other parts of the range are less than 5 kilograms. Fishing cats weigh as much as 15 kilograms, but females of this species are much smaller than males.

The fishing cat and flat-headed cat eat fish and are closely associated with water; the leopard cat and Iriomote cat also are reported to fish. They all eat small mammals and supplement their diet with amphibians, reptiles, and birds. In addition, leopard cats and Iriomote cats eat insects, and fishing cats are reported to eat carrion. Rusty-spotted cats eat small mammals, amphibians, reptiles, and insects.

Caracal Lineage: Caracal and African Golden Cat, plus the Serval

The serval (*Leptailurus serval*) and the caracal (*Caracal caracal*) look as though they could be closely related, but molecular biologists have determined that they probably diverged more than 10 million years ago. The divergence between caracals and African golden cats (*Profelis aurata*) is thought to have occurred in the late Miocene, more than 5 million years ago. Caracals and servals are elegant, relatively small cats with long legs. Caracals weigh from 8 to 20 kilograms; servals weigh about 10 kilograms; African golden cats vary from 8 to 16 kilograms. Most servals are spotted with markedly varying patterns, although many of these cats are black. They have large, untassled ears. The caracal is tan in color with large, black-backed ears that have long black tassels. The African golden cat is more robust in appearance, and the coloration of its spotting and background varies from red and yellow to smoky gray; some are without discernable spots on the sides of their bodies, but all have black blotches on their white or off-white belly fur. They have small black-backed ears. All three live in Africa but occupy different habitats. The African golden cat lives in rain forest and also in secondary or disturbed forests. The serval lives in moist savannas and woodlands. The caracal lives in dry thickets, savannas, and woodlands in Africa, but, unlike the others, extends into similar vegetation in Asia Minor and as far east as western India. All are solitary, terrestrial hunters that prey on small mammals, supplementing with birds, amphibians and reptiles, and insects. The caracal also feeds on hares, and servals sometimes fish. The caracal and African golden cat also hunt small ungulates, especially duikers. All three share their habitats with the larger leopard, and the caracal and the serval hunt in habitats frequented by lions and cheetahs.

Bay Cat Lineage: Bay Cat and Asian Golden Cat

At less than 10 kilograms, the bay cat (*Catopuma badia*) is smaller than the Asian golden cat (*C. temminckii*), which weighs from 10 to 15 kilograms. The bay cat has two color phases, reddish and blackish-gray, and the end half of the tail is conspicuously white underneath. The coat color of the Asian golden cat may be red, brown, or gray. The amount of spotting varies among individuals in both species, and both have distinctive faces with white lines bordered in black across each cheek. The bay cat occupies rain forest; the Asian golden cat ranges from rain forest into moist temperate forest. Both are thought to be primarily terrestrial hunters of mostly small mammals, which they supplement with birds, but we have no information on their food habitats beyond a few incidental observations.

Puma Lineage: Cheetah, Puma, and Jaguarundi

The puma (*Puma concolor*) and the cheetah (*Acinonyx jubatus*) diverged from a common ancestor nearly 10 million years ago; the jaguarundi (*Puma yagouaroundi*) and puma diverged more than 5 million years ago. The lineage probably originated in Eurasia with a subsequent dispersal of both the cheetah ancestor and the puma and jaguarundi ancestor into North America, where the last two species diverged and then expanded their ranges into South America after the Panamanian land bridge formed about 4 million years ago. In South America, the puma's and jaguarundi's range and habitat use overlap those of nearly all the cats in the Leopardus Lineage, with the exception of the Andean mountain cat. The cheetah's ancestor also dispersed into Africa. In western North America the speedy pronghorn antelope probably evolved in response to the cheetah's predation. The cheetah and the puma disappeared from North America at the end of the Pleistocene, about 10,000 years ago, along with many other cat species. Genetic evidence indicates that the puma recolonized North America from South America sometime later. The cheetah and puma are medium-sized cats weighing less than 80 kilograms; the jaguarundi is a small cat weighing about 5 kilograms.

The solitary jaguarundi of today lives from northern Mexico to northern Argentina in a great diversity of habitats ranging from thorn forest to wetlands and tropical forest. It hunts on the ground and also apparently in trees and eats small mammals supplemented with birds, amphibians and reptiles, and fish. It is sometimes called the otter cat because of its mustelid-like shape. It has two color types: reddish or grayish, and neither is spotted.

The tawny-coated puma lives in a wide range of habitats from Patagonia to the Yukon. It needs cover for hunting. The largest pumas (males of 70 kilograms, and females of 45) are at the extreme ends of its long range; the ones that live in the tropics are about half as big. The solitary, terrestrial puma kills ungulates, including white-tailed and mule deer, elk, and bighorn sheep in the northern reaches of its range and guanacos in the southern reaches. Pumas eat white-tailed and brocket deer in Central and South America. They supplement these prey with larger rodents and other mammals seasonally. In Patagonia, they prey on the introduced European hare. Through much of the central part of its range, the puma overlaps with the larger jaguar. Biologists have found great overlap in what they hunt as well, with white-tailed deer being the most important prey. But pumas eat more different kinds of prey, from snakes to coatis, than jaguars do.

The cheetah is the only cat that chases its prey over relatively long distances. Now it lives only in sub-Saharan Africa and in small relict populations in Iran and in the central Sahara. Once it lived through southern Asia into India. Among the cats, cheetahs are also unique in that females live alone, but males live in small groups called coalitions. Their primary prey are gazelles, but they also take hares and larger ungulates from time to time. Cheetahs live in grasslands and in dry savannas and woodlands. They do not eat carrion, and they lose many of their kills to larger predators, including lions and spotted hyenas. Nearly everywhere they live, lions dominate cheetahs.

Felis Lineage: Wildcat, Domestic Cat, Chinese Mountain Cat, Sand Cat, Jungle Cat, and Black-footed Cat, plus the Pallas' Cat

The Pallas' cat (*Felis manul*) and the Felis Lineage, also referred to as the Domestic Cat Lineage, diverged from a common ancestor 10 million years ago; the Felis Lineage radiated in the Pliocene (5 to 1.8 million years ago). The black-footed cat (*F. nigripes*) is the next oldest radiation in the lineage, followed by the ancestor of the wildcat (*F. silvestris*) and Chinese mountain cat (*F. bieti*) and the ancestor of the jungle cat (*F. chaus*) and sand cat (*F. margarita*). These ancestors probably evolved in isolation from each other, with an east-west split. The domestic cat (*F. catus*) was derived from the wildcat about 7,000 years ago, probably in northern Africa (see *When and Where Were Cats Domesticated?*).

All are primarily terrestrial hunters, and all the wild species are solitary; feral domestic cats sometimes live in groups. All the wild species are either African or Eurasian in their distribution; the domestic cat lives nearly everywhere. There is considerable interbreeding between feral domestic cats and wildcats in Europe.

The black-footed cat and sand cat are extremely small cats, weighing less than 3 kilograms. The sand cat is tan in color and stocky, with the ears set wide on the sides of its head. The black-footed cat has large black or brown spots on a background coat, which varies in color from cinnamon-buff to tawny. The rest are small cats, weighing less than 10 kilograms at most and usually less than 5 kilograms. The northern wildcats are similar in color to domestic tabby cats. The wildcat in Africa has a shorter plain coat and slightly longer legs than the northern form. The Chinese mountain cat is gray with long hair. The jungle cat has a plain, tawny, short coat and appears to have longer legs than the northern cats in this lineage.

The sand cat and black-footed cat live in arid environments, although they do not overlap in distribution: the sand cat lives in true sand deserts; the black-footed cat lives in arid scrub and scrub steppe. In Africa, the wildcat lives nearly everywhere except in rain forest and true desert, although it does require cover. It lives in Mediterranean scrublands and in the temperate forests of Eurasia. In Scotland it lives in heather moorland and even in bogs. The jungle cat lives nearly everywhere except in the jungle, occupying disturbed areas, dry-forest shrub, and reed beds along rivers from Egypt to Thailand. The Chinese mountain cat inhabits mountain terrain, up to more than 4,000 meters in elevation, with brush, forest, and steppe, from central to western China, including the steppe and steep slopes of the Tibetan plateau. All the cats in this lineage eat various species of small mammals, supplemented with birds, amphibians, and reptiles, depending on the particular habitats. Occasionally they kill an ungulate fawn and feed on carrion.

Pallas' cats have pale gray, thick, long fur, short legs, and live in the harsh continental-climate areas of central Asia from the Caspian Sea to central China up to elevations of 4,000 meters. While adapted to live in cold and hostile climates, they do not live where snow accumulates. They don't live in true desert, but they do live in semideserts, in hilly areas, and in steppes with rocky outcrops. Russian biologists report that, apparently unique among cats, Pallas' cats put on extra layers of fat when prey is abundant, to tide them over during long periods of inclement weather, when prey is in short supply. They live on pikas, small lagomorphs related to rabbits, supplemented by small rodents, partridges, and hares.

Leopardus Lineage: Ocelot, Margay, Oncilla, Chilean Pampas Cat, Pantanal Cat, Argentinean Pampas Cat, Kodkod, Geoffroy's Cat, and Andean Mountain Cat

The origins of the Leopardus (or Ocelot) Lineage extend back more than 5 million years in North America. The members of this lineage entered South America via the Panamanian land bridge, thought to have formed between 4.1 and 2.7 million years ago. Nine species are now recognized in this radiation. This includes the recent decision by biologists that the pampas cat should now be recognized as three species: the Pantanal cat (*Leopardus braccatus*), the Chilean pampas cat (*L. colocolo*), and the Argentinean pampas cat (*L. pajeros*). Experts expect that the oncilla (*L. tigrinus*) will also be recognized as two distinct species, Central American and Brazilian. All the cats in this lineage are small (weighing less than 20 kilograms), and the oncilla, kodkod (*L. guigna*), and some Geoffroy's cats (*L. geoffroyi*) are extremely small (weighing less than 3 kilograms). It is not surprising that this lineage holds no medium-sized or large forms, because the puma and jaguar occupy these ecospaces in its range. The cats in this lineage eat small rodents, supplemented with birds, snakes, lizards, and other small animals. Spot patterns of the species in this lineage can be so similar as to confuse experienced observers, but there is much intraspecific variation in background color and spot patterns.

These are all solitary cats, primarily nocturnal in their hunting, but radio-tracking data on some species indicate they may have crepuscular and even diurnal hunting bouts. The margay and oncilla are primary arboreal hunters but also hunt on the ground. The remainder of the species in this lineage are primarily terrestrial hunters. The oncilla lives in dry and humid evergreen forests and scrubland from southern Costa Rica to northern Argentina. Its distribution in the Amazon rain forest is not known in detail, but it usually lives at higher elevations than the margay (*L. wiedii*) or ocelot (*L. pardalis*). Margay and ocelot ranges extend from northern coastal Mexico to northern Argentina, where the margay is confined to humid tropical forests, premontane moist forests, and cloud forests, and the ocelot lives in a wide variety of habitats from rain forest to thorn shrub. A small population of ocelots also still lives in the Rio Grande Valley in Texas. Margays seem not to tolerate human disturbance, but ocelots can live in disturbed areas.

The Andean mountain cat (*L. jacobitus*) lives in open landscapes at high elevations in the central Andes. The Argentinean pampas cat (*L. pajeros*) lives in the high steppe on the eastern slope of the Andes from Ecuador to Patagonia; in Argentina, it lives in lowland steppe, shrubland, and dry forest habitats. The Pantanal cat lives in the humid, wet, and warm grassland and forest areas of moderate elevation in Brazil, Paraguay, and Uruguay. The Chilean pampas cat lives in subtropical forest and in high-elevation steppe in Chile on the western slope of the Andes. The kodkod lives in the southern conifer and deciduous forests of Chile and Argentina. The Geoffroy's cat lives in open woodlands, brushy areas, open savannas, and marshes, but not in subtropical rain forest or southern conifer forests, where it is replaced by the kodkod.

The Andean mountain cat, ocelot, and margay form one branch in this lineage, with the Andean mountain cat separating first, followed by the separation of the margay and ocelot. The second radiation was the pampas cats, which diversified in the Pleistocene. The third radiation within the lineage was the Geoffroy's cat, kodkod, and oncilla. The divergence in the three main groups probably occurred in North America, before the formation of the Panamanian land bridge. The radiations continued after these species entered South America. Today, the Andean mountain cat lives apart from any of the others in the lineage. The geographic ranges of the ocelot, margay, and oncilla overlap to a considerable extent. The Geoffroy's cat and kodkod in southern South America overlap to some extent but are reported to live in different vegetation types. The geographic ranges of the ocelot, margay, Pantanal cat, oncilla, and Geoffroy's cat overlap in northern Argentina and Paraguay. A natural hybrid has been reported between an oncilla and Pantanal cat.

HOW MANY DIFFERENT CAT SPECIES CAN LIVE IN ONE AREA?

We don't know all of the factors that control how many species can coexist in a given place, but some general patterns in species diversity have been discovered. In areas of comparable size, more species live in the tropics, and the number declines as one goes north or south toward the poles. The more stressed the environment, the fewer species will live there; for example, deserts and the Arctic tundra have fewer species than the wet tropical forests. The more favorable conditions are for biological production (in terms of the amount of plant and animal biomass produced annually), such as warm temperatures and abundant rainfall, the more species will be found there. Strongly seasonal environments have fewer species than aseasonal environments. There is a strong relationship between species diversity and habitat complexity. Mountains have greater species richness than flatlands. Natural disturbance also influences species richness. Landscapes disturbed by fire, floods, landslips, and similar events are a mosaic of different habitats that support a greater number of species.

The larger the area you consider, of course, the more species will live there, because by expanding the area under consideration, you also increase the chance of including additional habitat types and species whose ranges are restricted. This relationship is called the species-area curve. If you pick as your starting area, for example, the 2,168-square-kilometer Khao Yai National Park rain forest in central Thailand, you will find that four wild cat species are known to live there: the tiger, the leopard cat, the marbled cat, and the clouded leopard. We don't know why the Asian golden cat and fishing cat are not in the Khao Yai (in other areas they live in rain forest), but by expanding the area under consideration to the 514,000 square kilometers that make up all of Thailand, these two species are added, plus the jungle cat, the flat-headed cat, and the leopard, for a total of nine cat species.

Jungle cats are common in Thailand, but they live in disturbed areas around villages and in scrubby vegetation, perhaps because the dominant rain forest cats, which exclude the jungle cat from their habitat, do not find favorable conditions for themselves in the scrubby areas. We suspect that the leopard is excluded from Khao Yai because tigers live there, although leopards and tigers both live in rain forest on the Malay Peninsula. The flat-headed cat is restricted in its distribution to below the Isthmus of Kra at the southern end of Thailand and does not occur in northern Thailand. If you expand the area to include all of Southeast Asia, including the Greater Sunda Islands, you will find a total of 10 cat species. The only additional species is the bay cat, endemic to Borneo.

The number of species tends to increase with the size of islands as well. There are seven cat species on 427,300-square-kilometer Sumatra, four cat species on 126,700-square-kilometer Java, and two cat species on the adjacent 5,500-square-kilometer Bali; although the tiger is only recently extinct on Java and Bali. Also, the more remote the island or the longer it has been separated from the mainland, the fewer the species that exist there. Borneo is much larger (743,200 square kilometers) than the other Greater Sunda Islands listed above, but it has been separated longer and has only five cat species.

Geography matters in other ways as well. Through the shifting of the Earth's plates, mountains rise, creating barriers to dispersal. The formation of glaciers created barriers to dispersal in some parts of the world but also lowered sea levels in other parts of the world, creating dispersal corridors. This is how cats representing the different lineages dispersed from their places of origin to the different continents or ended up in one corner of their former range. For example, the entire Leopardus Lineage radiated just before and right after entering South America when the land bridge formed between North and South America (4.1 to 2.7 million years ago). The process of barriers appearing and disappearing is ongoing. Pleistocene (1.8 million to 10,000 years ago) cooling and glaciation caused massive extinctions and displaced many communities. Now, cats representing the different lineages have, in some cases, come to live together in one place or returned to old haunts in different forms.

Ten different wild cat species belonging to five different lineages live in the mountains at the eastern edge of the Tibetan plateau, in central China—one of Earth's most important biodiversity hot spots but also one with a large human footprint. Animals adapted to five different broad zones—northwestern mountains, western escarpments, Tibetan plateau, central basin, and subtropical forest—converge here. In these mountains or at their western or northwestern edge, are the Pallas' cat and Chinese mountain cat from the Felis Lineage, the only species that are found here but nowhere else; and on the southern edge are the jungle cat, also from the Felis Lineage; the Asian golden cat from the Bay Cat Lineage; the leopard cat from the Leopard Cat Lineage; the Eurasian lynx from the Lynx Lineage; and the clouded leopard, leopard, tiger (now extinct), and snow leopard from the Panthera Lineage.

Usually, all of these species would not be found within the same 10 square kilometers or even the same 1,000 square kilometers, but this mountain system is a very heterogeneous environment that provides living opportunities for many different cat species. Elevations range from 500 meters in the Sichuan Basin to snow-clad peaks of 7,000 meters. Habitats range from rock and snow to alpine meadow and thicket, to subalpine coniferous forest, coniferous and deciduous broadleaf forest, evergreen and deciduous broadleaf forest, and evergreen broadleaf forest. Riparian habitats border streams, rivers, and mountain lakes. The climate varies from maritime, with heavy summer rains and cool winters, to severely continental, with very cold winters and dry summers.

Ten cat species also live in Southeast Asia (as discussed above), representing four lineages along with the marbled cat of ambiguous lineage. Some of the species are found both here and in the Chinese mountain complex, and some are found only here. Three cat species are endemic to Southeast Asia—the flat-headed cat, the bay cat, and the marbled cat—and we assume evolved there. The fishing cat probably also evolved in Southeast Asia but now has a wider distribution into south Asia. These four species do not seem to enter the Chinese mountain complex. The Eurasian lynx, snow leopard, Pallas' cat, and Chinese mountain cat are not found in Southeast Asia. The tiger, leopard, Asian golden cat, and leopard cat live in both.

The pattern of species richness reflects the balance between regional processes that add species to communities (species formation and geographic dispersal opportunities) and local processes that contribute to local extinction (predation, competitive exclusion, adaptation, and chance events). Landscapes include a multitude of partly overlapping cat distributions, and the distributional boundary of each species reflects its relation with the environment. This includes the distribution of its potential prey, its competitors, the depletion of its potential prey by its competitors, its predators and pathogens, and its physiological response to the environment.

On a more local level, competition among cats determines how many may live in an area. No two species can occupy exactly the same niche—the place in an ecosystem where a species lives, what it eats, its foraging routes, activity times, and so forth. Coexisting cat species may avoid competition by hunting different prey, or by using different spaces, or by hunting at different times. Or they may be different enough in size or some other feature that they don't directly compete.

There are only three species—puma, bobcat, and Canada lynx—living in the Salmon River Mountains in Idaho. Pumas, the largest, dominate and are the most wide ranging, killing mule deer and elk in the winter and expanding their diet to include marmots and ground squirrels in summer, when they move to higher elevations. In winter, bobcats live in the bluffs and along the lower streams, where they kill mostly rodents, the occasional grouse, small birds, and doe and fawn mule deer. They take about the same fare in summer but at higher elevations. Canada lynx, about the same size as bobcats here, are rare and are confined to the upper elevations, even in winter, where they specialize in hunting snowshoe hares and red squirrels. With their large footpads, lynx sink less in snow than bobcats and pumas do. This snow-sinking factor is called foot-loading and is measured by grams of body weight per square centimeter of foot pads. Snowshoe hares have foot-loading values of about 10; lynx, about 100; bobcats, 300; and pumas, 1,000. These cats are separated by body size, locomotion adaptations, prey type, and habitat type, but, even with this ecological and morphological separation, the primary cause of death in bobcats is puma predation.

Tigers, leopards, fishing cats, leopard cats, and jungle cats live in Nepal's Royal Chitwan National Park. Conservation biologist Dave Smith radio-tracked fishing cats and found they do not go into the riparian forest but stick to riverbanks and the tall grassland and banks of old oxbow lakes beyond the forest. They fish and also kill rodents. Jungle cats live at the edge of the forest and are seen in the grassy areas near occupied and abandoned villages. They presumably live on rodents and the wayward chicken they can catch. Leopard cats seem to be restricted to the mixed forest. John found that leopards and tigers use the same areas of forest and grassland edges and kill the same ungulate prey species, but the smaller leopards kill prey that are about one-fourth the size, on average, of the tiger's prey (Seidensticker 1976). The leopard is more active in the day than the tiger and uses different trails and stream crossings from those of the tiger. Tigers kill leopards; leopards kill leopard cats.

The tiger, clouded leopard, marbled cat, flat-headed cat, fishing cat, Asian golden cat, and leopard cat—seven species—can potentially live in the same areas in Sumatra; they all occur on the island. (In this cat assemblage, competitors apparently have squeezed the usually tenacious and generalist leopard out, because it does not occur on Sumatra.) No one has yet determined if all seven species actually live in one area or how those that do live in the same area manage to divide resources. We assume two factors: that the Sumatran rain forest provides a variety of different microenvironments that don't vary much over the year as well as a variety and abundance of potential prey, and that each cat species is a specialist in its own right. These factors may result in the denser packing of species in this tropical rain forest on the equator than in subtropical Nepal at 27° north latitude, with its five species, or in temperate Idaho at 45° north latitude, with three species.

.3.

CATS AND HUMANS

WHY ARE WILD CAT NUMBERS DECLINING?

Pulitzer Prize–winning biologist Jared Diamond refers to the mechanisms of recent species extinctions as the "arsenal of species extermination," or the "Evil Quartet": habitat fragmentation and destruction, overkill, effects of introduced species, and secondary extinctions (the last of which occur as a result of the extinction of another species or disruption of ecological processes). Another Pulitzer Prize–winning biologist, E. O. Wilson, refers to these as "the mindless horsemen of the evolutionary apocalypse." Extinction and human presence go hand in hand. Wild cats are in decline because they are being killed accidentally, such as by vehicles, or deliberately, for revenge or to supply illegal markets. The habitat they require is disappearing. The prey they need to survive are overexploited by people or are dying from introduced diseases. Most people aren't aware of or do not appreciate what wild cats need to survive. In many parts of the world, there are no effective programs for the protection of wildlife and their habitats in general or for cats specifically. Cats threaten people and their livestock, and the results are continuing confrontations and dead cats. We know that most, if not all, recent extinctions of animal species can be related directly or indirectly to human activity. If wild cats are to remain a vital part of our legacy, we must act to make it so.

Human activities have sent wild cat numbers into a rapid downward spiral. Although the decline of big cats such as tigers and cheetahs is fairly well known, many small cats are threatened too. Each year in South America, huge numbers of Geoffroy's cats are trapped and killed for their beautiful black-spotted fur.

It has only been since the 1970s, with the passage and implementation of the US Endangered Species Act, the Convention on International Trade in Endangered Species of Wild Flora and Fauna (CITES), and the identification of the threats to many cat species with the publication of the International Union for Conservation of Nature and Natural Resources (IUCN) (now the World Conservation Union) red data books, that the challenges presented by the loss of species have become more widely understood by the general public. Because conservation biologists are paying close attention, we know that one or more populations of every wild cat species are threatened or endangered.

HOW DO PEOPLE AFFECT WILD CATS?

Almost inevitably, conservationists posit that the continued rapid growth of human populations, combined with an increase in per capita consumption of resources, is placing unprecedented demand on the Earth's renewable and nonrenewable resources, from trees and soil to oil and diamonds, leading to increased threats to all endangered species, including wild cats. We are in an environmental crisis of our own making, and we are losing our biodiversity. The threat to the continued existence of wild cats is but one loud roaring reminder of our critical conservation issues.

Recently, biologists have been fine-tuning our understanding of the role of human population size and its relationship to the rising threat to birds and mammals. They have advanced our awareness by comparing the human population density of most of the world's countries with the proportion of species threatened there. A surprising finding is that only about one-third of the variation among the threat levels is directly attributable to the density of humans. That leaves about two-thirds of the variation in extinction threat to be explained by other factors. As Michael McKinney (2001), the author of this insightful and comprehensive report, has explained, other factors, such as the standard of living and government policies, can influence extinction threats in a nation, independent of its human population size. Traits of endangered birds and mammals themselves also make them prone to extinction.

Across the world, the human population is increasing faster than the proportion of species threatened. In other words, few people may do far greater damage in pristine systems through hunting and land clearing than many people do by living at high densities in urban environments. Even in large areas, a few people can do immense damage. Nowhere is this clearer than on islands, where people have catastrophic impacts, such as by releasing domestic cats, which feed on native species with devastating effect.

Many people blame human population growth for the decline of cats and other wildlife, but human choices will determine the ultimate fate of wild animals. As long as people choose to purchase spotted-cat furs, for instance, they will appear in markets until the last spotted cat is gone.

Without question, though, human population density is part of the problem. McKinney found that Asia contains the nations with the highest proportions of threatened species and the highest number of people. Africa and Latin America contain many nations with small human populations and a relatively low proportion of threat to their mammal and bird species. However, John has worked on cat conservation issues in many Asian countries and cannot imagine telling a local official, "The problem here is that you have too many people." The critical issue is what people do, not simply that they exist. And what people do is strongly influenced by their standard of living, government policies, long-held cultural traditions, and what they value in general. For example, deforestation and other kinds of habitat loss, a major cause of extinctions, are directly correlated to human population density, but effective government policy can stabilize the amount of forest cover retained and even restore what has been lost.

WHAT IS THE LARGE CARNIVORE PROBLEM?

In a crowded world, large carnivores are the most challenging faunal group to conserve. Conservation biologists call this the large carnivore problem. However, enlightened government and management policies can offset the impact of high human densities. Large carnivore populations can increase after favorable legislation is implemented, despite further increases in human population density. The puma in North America and the Eurasian lynx are examples of species that have benefited from strong positive policy initiatives. This finding is encouraging for conservation biologists as they devise strategies to protect and restore cat populations in other areas of the world, where, historically, policies have not been large-carnivore friendly.

Leopards, like all large carnivores, are among the most difficult animals to conserve in our crowded world. They require large areas with plenty of prey (prey that people also find good to eat) as well as a certain amount of solitude for females to raise young, and they inspire equal amounts of fear and admiration among the people who share their habitats. However, strong conservation policies and programs, such as those in North America and Europe that have enabled the recovery of pumas and Eurasian lynx from the brink of extinction, have shown us that a future for large carnivores is possible.

WHAT IS HAPPENING TO ALL THE WILD CAT HABITAT?

Habitat destruction, including its degradation and fragmentation, is the most pervasive threat to endangered species. This destruction results from several different activities: the conversion of land to agriculture or the adoption of certain agricultural practices, including livestock grazing; mining and oil and gas exploration and development; logging; infrastructure development, including road construction to support logging and mining activities; military activity; outdoor recreation, including off-road vehicle use; water developments; pollutants; land conversion for urban and commercial development; and the disruption of fire ecology including fire suppression. The human footprint is huge, and it covers the globe.

Conservation organizations, including the Smithsonian's National Zoological Park, the Wildlife Conservation Society, Conservation International, and the World Wildlife Fund, have been mapping the human footprint and searching for "the last of the wild," that is, areas little affected by human activities where we might save wildlife. As you might expect, some biomes have more and larger wild areas than others. More than 67 percent of the area of North American tundra is wild, but no wild cats live there. The wildest areas of Old World tropical and subtropical moist broadleaf forest are all in China, where as many as 10 different wild cat species live. But wild areas encompass less that 0.03 percent of that biome. Large areas of Alaska and northern Canada and Russia are still wild, and these are home to Canada and Eurasian lynxes and the puma in western North America. Vast stretches of the African Sahara and the Arabian Peninsula are wild; this is sand cat country and, once and potentially again, cheetah country. The once vast and foreboding rain forests of central Africa have been carved up, but some large wild tracts remain, home to the African golden cat and the leopard. Large areas of the Tibetan plateau and its mountain ranges, home of the snow leopard, are wild, as are the deserts of central Asia, where the Pallas' cat lives at the edge. Some wild rain forest remains on Borneo, which might protect the bay cat and leopard cat. The Amazon Basin still has considerable wild lands where the jaguar, puma, jaguarundi, ocelot, margay, and oncilla live. But we cannot rely on the "last of the wild" to save wild cats. Rather, we have to figure out how to create conditions and processes that let wild cats live in human-dominated landscapes.

Because wild cats are at the top of the energy pyramid, human alteration of food webs is a primary, but little appreciated, threat to cats. People appropriate more than 40 percent of the net primary productivity—the green material—produced on Earth each year, with a profound effect on most natural food webs. People appropriate resources when their livestock graze on plants that wild ungulates need to survive. People also kill the large cats' primary food, large ungulates, to feed themselves and to keep those ungulates from eating their crops. Vast areas that are potential habitat for tigers, for example, are devoid of prey through much of the tiger's range, from the Russian Far East through much of south and Southeast Asia, where less than 20 percent of the potential habitat is protected in reserves. Hunters kill European rabbits, reducing the primary prey of the critically endangered Iberian lynx. Inevitably, hunters complain that carnivores take what they consider to be their "game." In the past, governments put bounties on the heads of carnivores, and poisoning, trapping, other means of killing them have been devastatingly effective.

Today, conservation biologists try to work with hunters and devise coexistence recipes based on understanding and tolerance.

Various populations within wild cat species were once broadly linked by favorable habitat so that cats could move among them. Habitat fragmentation, in which human activities cut the linkages, divides wild cat populations into smaller and more isolated units; cats live on habitat islands in a sea of humanity. In a metapopulation, which is a set of local populations connected by dispersing individuals, the movement of those individuals is highly restricted because of the hostile conditions between habitat patches. Wild cats living in small habitat islands cannot move through the hostile human landscape to reach other areas, where other individuals of the same species may live. Neither can they move to colonize another area of habitat. Wild cats tend to become extinct in these small patches. A disease epidemic, a rash of poachers, an upsurge in predators, or a catastrophe such as a flood or fire may wipe out all of the cats in the patch. Or, simply by chance, all the young produced in a very small population are either males or females, or no young are produced at all. Genetic deterioration due to inbreeding is a further threat. Moreover, once the cats go extinct in a patch, it cannot be recolonized because of the loss of all the corridors and links to where other wild cats of the same species remain. Conservation biologists call this the extinction vortex.

WHY ARE SOME CATS MORE VULNERABLE THAN OTHERS TO THE ACTIONS OF PEOPLE?

Species with large body size, low abundance, low reproduction rates, specialized habitat requirements, and isolated populations are thought to be most vulnerable to extinction. Large carnivores with large home ranges, such as tigers and lions, are prone to local extinction because their ranges make them liable to move outside their protected areas and come into conflict with humans. To protect these species, land-use policies must include not just reserves but also large buffer zones around reserves, where these animals are tolerated by people and also receive good protection. The persistence of carnivore populations depends on their individuals' abilities to disperse through various habitats that offer less than ideal conditions. Tigers, which evolved in forested landscapes, are poor dispersers in the landscape mosaics they live in today, because they do not like to cross open areas created by agricultural activities. Leopards, on the other hand, are masters at dispersing across the most desolate habitats, making them less vulnerable to human activities. Large-bodied carnivores such lions and tigers are more prone to local extinctions, whereas smaller cats such as leopards and servals are more likely to persist because of their dispersal abilities. Many smaller cat species also seem to tolerate living among

Wild places are disappearing, so ways must be found to enable cats to survive in human-dominated landscapes. Setting aside large protected areas, bans and limits on hunting of both predator and prey, growing appreciation of the needs and values of wildlife, and the increasing movement of people from rural to urban environments have contributed to the recovery of pumas in the western United States.

people. Fishing cats live next to the airport in Colombo, the capital of Sri Lanka, and jungle cats and leopard cats commonly live near Asian villages.

Living in very large remote areas isolated from human disturbance helps cat populations to survive, but there are few such areas. To compensate for the inadequate size, isolation, and remoteness of reserves, conservation biologists are trying to maintain or restore habitat connections between them, so that cats can move from one to another. Otherwise, cats attempting to disperse from the reserves are simply lost between them in the "sinks" created by human conditions.

Wild cats' ecological adaptability, dispersal efficiency, and tolerance of human activity are the traits that determine how they will respond to the human-dominated environments that are the norm today. Biologists simply do not know or have not been able to define the tolerance of most wild cat species, and this information is essential to designing management plans that can lead to recovery of their populations.

HOW DOES THE WILDLIFE TRADE AFFECT CATS?

Changing economic and political conditions affect wild cats in different ways. When economic conditions worsened after the fall of the Soviet Union in the early 1990s, the population of Amur tigers that had been slowly recovering took a nose dive because of intense poaching. The change in the political system relaxed protection and, at the same time, created access to other markets, especially along the newly opened border with China. Poachers took advantage of this, just as they continue to take advantage of social and economic turmoil in many other countries.

In a more positive example, western North America has changed from an economic base of predominately rural farming, ranching, logging, and mining to one more dependent on industries such as financial services and computer and information technology. More people live in towns, and they tend to use wild lands for recreation rather than resource extraction. These "new westerners" view wildlife, especially the large carnivores, in a positive light, as something that adds value to their life. Here people are encountering pumas directly and are learning to live with them. The remaining ranchers, on the other hand, view these same animals in a negative light because they threaten the ranchers' livestock and their already impaired livelihoods. These tensions have to be carefully negotiated in the political process. So far, the balance is in favor of the puma and the wolf, whose populations are now much larger than they were a century ago. A negative effect of this economic shift, however, is that large landholdings are being divided into small "ranchettes," which fragment habitats, or developed as subdivisions, which intrude into the winter range of deer and elk.

In south Asia, many protected wildlife areas, including tiger reserves, have substantial populations of people living in them. But people are increasingly willing to move out of reserves to places where they have access to life-changing benefits such as schools, better health care, and jobs, and this benefits the wild cats in these areas. Conservation biologists are learning that protecting wildlife can often best be achieved by providing rural people with new job skills, including those that allow them to join the computer age.

People also kill cats and other wildlife to sell their parts and products as food and medicine. Most of this trade is illegal across international boundaries, but the laws haven't stopped it. Furthermore, *all* the countries involved in it, whether they are where the wildlife live and are hunted or they are distant countries where the trade

Increasing affluence in Asia, as well as large, relatively wealthy Asian communities in North America and Europe, have fueled an increase in the demand for traditional medicines that incorporate tiger bone and other wildlife parts and products. Curtailing the illegal wildlife trade through education and law enforcement is essential. (Photograph by Judy Mills)

takes place, are having an increasing impact on the wildlife. For example, until recently, the largest and fastest growing markets for medicinal products reputed to contain tiger bone were the United States, Canada, and western Europe. Relatedly, a number of "tiger farmers" in the United States have been recently convicted for raising tigers for their hides and to sell the meat to specialty shops.

Wild cat parts and products are used both as part of traditional medicine systems and as food. In the past, the cost of products purported to contain wild cat parts generally placed them beyond the reach of the average person. As Asian economies grow, however, and individuals accumulate more wealth, their buying power increases. With an increase in disposable income, these products become affordable, thus increasing the market demand and resulting in more commercial killing to supply that demand. Moreover, in the consumption of these products, the distinction between food and medicine becomes blurred; many think whatever is good for you in small amounts is better for you in large amounts. So rather than a taking a pinch of tiger bone to treat an ailment, the newly rich indulge in a dinner of tiger penis soup, which is also a status symbol. This behavior is leading to the wholesale destruction of all wildlife in the tropical forests throughout Southeast Asia. As the actor Jackie Chan and other celebrities point out in the public service announcements broadcast on television in many Southeast Asian countries and in China: "When the buying stops, the killing will too." And that is conservation's premiere challenge: Stop the buying. Get rid of the demand, and the rest can follow.

WHAT ARE THE MOST ENDANGERED CATS?

The most endangered cat, not to be confused with rare, is the Iberian lynx. A rare animal is not necessarily an endangered one; an animal that is currently endangered may once have been quite widespread and common. The Iberian lynx is such an animal, once widespread in Spain and Portugal. It was regularly hunted and considered by some to be vermin. Formerly thought to be a subspecies of the Eurasian lynx, the Iberian lynx has now been shown to be a separate species. But even though it is arguably the best-studied small cat in the world, it is a species we may see go extinct in our lifetime.

The first blow to this species came when the poxvirus myxomatosis was introduced from South America in the 1950s, and European rabbits, which had no natural immunity, were decimated. Rabbits form most of the Iberian lynx's diet, so their decline hit these cats hard. Just as the rabbits were developing some immunity and beginning to recover, a new virus, hemorrhagic pneumonia, hit in the late 1980s, again causing a high mortality in adult rabbits. These diseases that killed the Iberian lynx's prey, along with continued poaching, being killed by cars as these cats try to cross highways, and land-use changes that do not favor the lynx or its prey, are reasons this species may not survive. If that is not enough, a case of bovine tuberculosis was recently found in a dying lynx, the first ever such case in a wild cat. Populations of red deer, fallow deer, wild swine, and feral cattle live within the habitat of the lynx in Spain's Doñana National Park, and these species could be reservoirs for the disease.

Only two populations of this lynx remain in Spain, one in Doñana National Park and the other in the Sierra Morena. Another population may remain in Portugal. Each population consists of only a few tens of individuals.

No cat species that we know of has gone extinct within historical time, but many populations or subspecies have been extirpated. The tiger subspecies that lived on the Indonesian islands of Bali and Java, those that lived around the Caspian Sea, and the wild tiger of central China are gone. The eastern puma is gone except for a remnant population in Florida, called the Florida panther. The fishing cats that lived on the delta of the Indus River are gone. The South African Cape lion is gone, and so are the lions, leopards, cheetahs, and servals that lived north of the Sahara. The list goes on. Basically, only domestic cats, both feral and pets, are secure.

Who decides if a wild cat species is endangered or threatened with extinction? Most countries have their own endangered species laws today, each country with its own definition. Internationally, we rely on the World Conservation Union's 2003 IUCN Red List of Threatened Species. Unlike the US Endangered Species List, the Red List does not have the force of law, but its categorization of species, from extinct to critically endangered to being of least concern, is considered scientifically authoritative.

CITES, or the Convention on International Trade in Endangered Species of Wild Flora and Fauna, was established in 1975 to ensure that trade in wildlife species is managed for their sustainability. As participants in this treaty, individual countries sign on as parties to the convention and agree to abide by its rules. Species are listed in CITES appendix I, II, or III: appendix I is the highest category of threat, and trade in these species, which threatens them with extinction, is severely restricted; trade in species listed in appendix II must be based on sustainable harvests. In appendix III, countries can list the protected native species that live within their borders, in order to prevent or at least restrict their exploitation.

The Red List includes one critically endangered species of wild cats, which means it has a high probability of extinction within five years or two generations, whichever is longer. Five more are listed as endangered, 13 as vulnerable, 7 as near-threatened, and 13 as those of least concern. There are 21 wild cat taxa in CITES appendix I and 20 in appendix II. Some species listed on the IUCN Red List are listed only in appendix II of CITES (the bay cat is an example), whereas other species (such as the caracal) are listed as species of least concern yet appear in

Conservationists fear that the Iberian lynx may be first cat to go extinct since the saber-toothed *Smilodon* did about 10,000 years ago. This cat is beset by nearly everything that can threaten a species: small distribution in Spain and Portugal, continuing loss and fragmentation of habitat, hunting pressure, the decimation of its specialized diet of rabbits by introduced diseases, introduced disease killing the lynx themselves, and high mortality as roadkills. (Photograph by John Seidensticker)

CITES appendix I. The reason for this disparity is the narrow scope of CITES's focus on trade. Although the bay bat is considered endangered, it is not clear that this species is affected by the wildlife trade, but the caracal is or may be affected by it. These species are shifted from one category to another all the time, with the greatest number of shifts from appendix II to appendix I.

Appendix 2 in this book lists the status of each wild cat species on the 2003 IUCN Red List, the US Endangered Species List, and CITES. The threat categories of wild cats in the Red List and the US Endangered Species List contain some differences. The US list generally puts species in a higher threat category than the Red List. Some of this disparity is the result of insufficient information; some, the result of the lengthy review process necessary to change species from one category to another; and some, the result of genuine disagreement among biologists or governments. All would agree, however, that any wild cat species at any risk of extinction is one too many.

WHAT DO WE MEAN WHEN WE SAY CATS ARE UMBRELLA, FLAGSHIP, LANDSCAPE, KEYSTONE, OR INDICATOR SPECIES?

In regard to animal species, the words *umbrella*, *flagship*, *landscape*, *keystone*, and *indicator* have distinct but overlapping meanings. Conservationists use various wild cat species as flagship, umbrella, and landscape species, and when they influence ecosystem processes, biologists call them keystone species or indicator species.

Flagship species are popular, charismatic species that serve as symbols and rallying points to stimulate conservation awareness and action. Flagship species are about marketing; for instance, tigers make a good flagship species because people value them so highly—more than, say, skunks. People have inherently strong feeling about cats, which makes them useful as flagship species for conservation biologists trying to increase awareness about threats to cats and life's diversity in general. Cats are charismatic; many people relate to them in positive ways, so some species have become symbols and leading elements of entire conservation campaigns. Appeals on behalf of charismatic species usually raise more money than those on behalf of less charismatic species. Tigers probably rival giant pandas as the top poster children for immense international and regional conservation activities. Organizations dedicated to securing a future for tigers include *tiger* in their name. Save The Tiger Fund, the Tigris Foundation, and the Tiger Foundation are examples. Other cats that have emerged as flagship species include the Florida panther, cheetah, lion, Iberian lynx, snow leopard, and jaguar.

The tiger, like all cats, is also an umbrella species. This means that its areas of occupancy and home ranges are large enough and its habitat requirements are broad enough, or exacting enough, that setting aside a sufficiently large area for its protection will automatically protect many other species that may also need large blocks of relatively natural or unaltered habitat to maintain viable populations. Protecting umbrella species may preserve genetic and ecological processes that maintain diversity. So the concept of umbrella species is a big one.

Tigers and other large cats are also landscape species, in that it takes the conservation of entire large regional landscapes to meet their ecological needs. The term *landscape species* combines the concept of flagship species, with all its charismatic content, and the concept of umbrella species, with its content of ecological processes,

The beauty and charisma of tigers have make them symbols of the conservation movement, helping to create support for efforts to secure a future in the wild for them and all other threatened but less appealing creatures. In addition to this role as a "flagship species," the tiger serves as an "umbrella species." Successful efforts to save tigers, with their need for huge, relatively undisturbed habitats, will protect other wildlife in those habitats from the threatening rain of adverse human activities.

both broadly spatial and species inclusive. The Terai Arc, which extends 1,000 kilometers across northern India and Nepal, is one such tiger landscape that conservationists are trying to create through a combination of protection and restoration.

Biologists studying intertidal invertebrate communities coined the keystone species concept. A keystone species is any that plays a vital role in a biological community. It is a species that through its size, or activity, or productivity has a greater impact on its community or ecosystem than is expected on the basis of its relative abundance. The real impact of a keystone species is usually detected when the species is no longer present. For example, in tropical wet forests, if tigers are absent, more leopards occur. These two species kill different prey—leopards usually kill smaller and more diverse species—so the effects of this shift cascade through the entire community.

Finally, there is the concept of the indicator species. These are species whose presence or absence or a change in their distribution and abundance reflects changes in environmental quality or other measures. For example, changes in the time that some plants flower or that some mammals give birth in seasonal environments may reveal that the climate has changed. Some species are sensitive to habitat fragmentation, pollution, or other stressors that degrade biodiversity, and monitoring their population changes may alert scientists to those environmental changes. Population trends among fishing cats and flat-headed cats may indicate the quality of water and streamside habitat. Indicator species are useful in management and public relations because they can signal an increased need to protect water quality, which may gather more public support than a call to protect the species itself, although the latter may be accomplished at the same time.

HOW CAN WE SAVE WILD CATS?

The answer to this question is contained in this quote from Stephen R. Humphrey and Bradley M. Stith: "Conservation of species and undamaged habitat as a practice of human culture has developed like a three-legged stool. Each leg is necessary but not sufficient. The legs of the conservation stool are sustainable use of renewable resources, species recovery, and habitat preservation. Conservation can progress by focusing on each of these, defining their limits, developing improvements, and preventing dysfunction" (1990, 341).

This is a powerful statement, a vision that encompasses just about all we have to do. However, after working on wild cat conservation for three decades, we have articulated another formula, the Five Cs of Conservation: For *carnivores* to have a future, there must be *core* protected areas in the form of reserves, habitat *corridors* connecting the core areas, and the participation and support of human *communities* that affect and are affected by the cores and corridors, and to make it all work, there must be effective *communication* among local communities, their local supporters, and well wishers everywhere. This extends rather than replaces the formula of Humphrey and Stith.

The traditional approach to securing a future for endangered species prescribed a simple formula: protect a few pieces of nature by keeping people out; leave them undisturbed; the animals and their habitats will survive there indefinitely. Conservation biologists now believe that this approach represents hospice ecology: watching carefully and compassionately as species inevitably slip into extinction.

Ultimately, human values will determine whether we sustain landscapes with wild cat species intact. The troubles that wild cats have stem from human conflicts over values, over what matters most to people. People will almost always put their own needs and the needs of their families and communities first, so somehow the presence of wild cats must help, or at least not hinder, people's ability to meet their own needs. To secure a future for wild cats, conservation actions must be adaptable, relevant, and made socially acceptable by linking the welfare of cats to that of people who live near them. A better future for all of us lies in establishing sustainable relationships between people and resources.

Legal protection and the establishment of protected areas have been cornerstones in programs to check declines and restore threatened and endangered wildlife, and such goals are accomplished only through tremendous effort by conservationists. Protected areas are important building blocks in wild cat conservation, but many wild cats live outside protected areas, move in and out of them, and must travel between them. Any formulation for the survival of wild cats has to include the protection of these individuals. Recognizing this, conservationists are searching for ways to partner with people living among wild cats, because they have the most to lose or gain.

In many areas, what can be accomplished through legal protection and reserve establishment seems to have reached a plateau, and we must find other ways to increase the population sizes of cat species of concern. These ways include securing and protecting larger reserves, protecting essential habitat outside reserves, incorporating the protection and habitat requirement of target species within land management systems surrounding reserves, and linking core reserves through connecting corridors.

Our best opportunities to increase the chances of survival for wild cat species are in those middle landscapes that lie between the urban and the wild. Because people are the dominant force driving what happens there, conservationists must engage these people in ways that turn this into a win-win situation for both them and the cats. This "local guardianship" concept is the key to securing the future for many wild cat species. We cannot simply proclaim that populations of many wild cats are at risk of extinction. As conservation people have said: No more prizes for predicting rain. Prizes only for building arks. We have to see a future for wild cats through restoration that improves living conditions for everyone. Wild cats can be stars in the ecological recovery that improves the lives of people who live near them.

In the mid-1970s, conservation practitioners tracking the wildlife trade noted that hundreds of thousands of wild cat skins were being shipped from source countries, mostly in the tropics, to user countries, mostly in western Europe and the United States. In large measure this was the impetus for CITES, the Convention on International Trade of Endangered Species of Wild Flora and Fauna. If wild cat species, especially those with spotted or striped fur, were to survive, this trade would have to be restricted or, better, stopped. CITES became the legal instrument for curbing the fur trade, but the real muscle came from a concentrated awareness campaign conducted by many conservation organizations. This campaign targeted those who bought coats and other items made of these cat skins, including fashion leaders and celebrities, pointing out that they personally were dooming cats to extinction. By and large, this awareness campaign worked and essentially eliminated much of this threat to wild cats. More recently, the use of wild cats and other wildlife as medicine and food is sparking grave concern, and conservation organizations are again working to use awareness to curtail this. As yet no downward trend in consumption has occurred, but the program is just coming together. The fate of many wild cats lies in its success or failure.

WHAT CAN I DO TO HELP?

The first thing you can do to help is to learn as much as you can about the conservation challenges facing the cat that you are particularly interested in. Reading this book has given you an overview of the conservation issues and the underlying biology of the cats upon which effective conservation actions must be based. Go to your local zoo and really watch the cats that live there. Compare them with each other. Ask zookeepers about the different species and individuals and their temperaments until you have more questions than they can answer. Seek answers to the remaining questions from other authorities and through extensive reading. Get to know the many conservation-minded nongovernment organizations (NGOs) that use cats as flagships. Find out just how each organization is addressing the conservation issues for the wild cat species that has caught your attention. Use your knowledge of wild cat biology and conservation issues to increase awareness among your family and friends.

One person can make a difference. Mark Baltz was a graduate student at the University of Missouri, Columbia, whose mascot is the Bengal tiger. He wrote an edi-

Income generation through ecotourism is one way to make the conservation of cats and other wildlife relevant and valuable to people who live near and among them. Everyone can support this, whether by taking an outing to a local national park or wildlife refuge or by traveling to India's famed Ranthambore National Park, where tigers and tourists mingle among the ruins of an ancient fortress.

torial for the local paper asking if the supporters of the University of Missouri should not also be interested in their mascot's continued survival in the wild. This caught the attention of Chancellor Richard Wallace. He met Baltz, and out of that meeting "Mizzou for Tigers" was born, to raise awareness of the plight of wild tigers, to raise money for fellowships and faculty appointments, and to support wild tiger biology and conservation programs, all linked to the school's Bengal tiger mascot. Hundreds of schools have chosen wild cats of one species or another as mascots, and similar actions there would be helpful to those cats.

Conservation-minded NGOs are always looking for monetary and human resources. Contribute where you can, both in dollars and by volunteering. Wild cat biologists are always looking for help—for example, volunteers to help maintain our paper filing systems and enter data into our electronic databases.

Go see a wild cat. Be an ecotourist or join a study tour in which volunteers help a wild cat biologist in the field. Ecotourism and study tours add value to living wild cats because they bring attention and money to local economies. You may never see a wild cat because of their nature, but your interest makes their lives matter, and that is an enormous contribution. The inspiration for many wild cat biologists and conservation-minded NGOs is the declaration of anthropologist Margaret Mead: "Never doubt that a small group of thoughtful, committed citizens can change the world. Indeed, it is the only thing that ever has."

WHEN AND WHERE WERE CATS DOMESTICATED?

Jared Diamond points out that "domesticable animals are all alike; every undomesticable animal is undomesticable in its own way" (1997, 157). In fact, wild cats possess many of the traits that have made other species undomesticable: They are solitary and territorial. They tend to have a nasty disposition. As carnivores, they don't efficiently convert food biomass into meat for us to eat. And they often don't breed well in captivity; surprisingly, most small wild cats breed relatively poorly in zoos.

Nonetheless, cats have been domesticated, or maybe not precisely. Diamond says domestication involves wild animals' being transformed into something more useful to humans. If, however, cats' primary use to us has been as killers of rodent pests, it would be difficult to imagine how we might have made a wildcat, already a supremely well specialized hunter of rats and mice, more useful. In fact, all that may have been necessary to begin the bond between people and cats was the proliferation of rodents around the granaries of the first farmers and some cats' willingness to tolerate the proximity to people in exchange for access to abundant prey.

Perhaps as a result, there are few morphological differences between domestic cats and their most likely wild ancestor, the African subspecies of the wildcat. So when archaeologists find cat and human remains together, as they did in a Jericho (modern Israel) site dating to 7000–6000 BCE and in an Indus Valley site dating to about 2000 BCE, it's not always clear whether the cats were domestics, or captives, or wildcats killed for fur or food.

The earliest clear sign of a familiar relationship between people and cats is a cat's jawbone excavated from one of the earliest human settlements on Cyprus, about 8,000 years ago. Because no cats occurred naturally on this Mediterranean island, the colonists who first arrived by sea must have brought cats with them. Bones of mice are also found here, perhaps less welcome stowaways on boats and another reason to include cats onboard. But even here, the cats may merely have been wildcats captured as kittens and tamed, a practice people the world over have indulged in.

Taming wild cats was formalized with cheetahs and, to a lesser extent, caracals in historical times. In the Middle East and India, cheetahs were captured, tamed, and trained to hunt gazelles beginning about 1,000 years ago and continuing into this century. Caracals were trained to hunt hares and birds. But cheetahs failed to breed in captivity until the 1950s, and thus domestication was impossible.

Specialists generally agree that African wildcats must have been first domesticated in Egypt between 4,000 and 5,000 years ago. Cats begin to figure in Egyptian art about 2000 BCE, or 4,000 years ago, with one exceptional image of a cat wearing a collar dating to about 2600 BCE. By about 1600 BCE, domestic cats clearly and frequently appear in art, shown sitting under chairs, eating fish, playing, and helping people hunt birds in the Nile Delta's papyrus swamps. From Egypt, domestic cats very slowly diffused throughout Europe and Asia.

African wildcats taking advantage of the rodents that thrived around human settlements in ancient Egypt gave rise to domestic cats that now live around the world. Scientists believe the cats were first domesticated between 4,000 and 5,000 years ago, but wildcats and people probably lived commensally for much longer.

HOW DID DOMESTIC CATS SPREAD FROM EGYPT TO THE REST OF THE WORLD?

The Egyptians tried to keep cats for themselves by making it illegal to export them to other countries. Their embargo seems to have been surprisingly effective, because there is little unequivocal evidence of domestic cats outside Egypt until 500 BCE in Greece, where a marble block depicts a leashed domestic cat squaring off with a leashed domestic dog. Earlier evidence of possibly domestic cats outside Egypt, including bone and foot prints dating back to 2500–2100 BCE from the Indus Valley's Harrappa culture in what is now Pakistan and western India, may represent local domestication or captive wild cats. Harrappan feline figurines depict cats with collars, but other figurines with collars are of rhinos and other species that were captive, not domestic. An ivory statuette of a cat from 1700 BCE Palestine and a fresco and sculptured head of a cat from 1500 to 1100 BCE Crete are better evidence of domestic cats outside Egypt, because both countries had trade connections with Egypt, but the scarcity of cat artifacts suggests that domestic cats remained confined to their center of domestication for millennia. And even when they did begin to appear more often in Europe, they attracted little fanfare; also, it is likely that cats were not actively traded but rather largely moved themselves, in the words of Konrad Lorenz, "from house to house, from village to village, until they gradually took possession of the whole continent" (1988, 19). It has also been suggested that the movement and expansion of cats followed that of the black rat, which was native to Asia, the brown rat, native to eastern Asia, and the house mouse, from southern Europe and Asia. Cats moved from Greece to Italy in the fifth century BCE and spread with the Romans through Europe via imperial routes.

Cats reached Britain by the fourth century CE and penetrated all of Europe and Asia by the tenth century. Surprisingly, cats weren't known as ratters and mousers until the fourth century CE in Rome—both Greeks and Roman employed domestic polecats and ferrets for this purpose—and Romans didn't even have a word to describe domestic cats specifically until then.

The movement of cats along established trade routes, including maritime routes, may owe more to human intervention than to their own agency. Cats were popular shipboard companions because they killed rodents and were considered good luck. Cats in this role reached islands and other far-flung parts of the world during the Age of Exploration, from the thirteenth through the eighteenth century, when Europeans sailed the world and began to settle in foreign lands. Few of even the most remote of uninhabited oceanic islands escaped colonization by domestic cats, which quickly went feral (see *Why Are Feral Cats—and Pet Cats—Sometimes a Menace?*). It is possible to show the genetic linkages between the domestic cats in various parts of the New World and the domestic cats in the countries from which the majority of European immigrants came. To this day, domestic cats in New England are fairly similar genetically, but the domestic cats of New York City stand out as distinctly different. Why? Because New York was first settled by the Dutch, whereas the rest of New England was settled by the English, and each group brought their own local cats with them.

WHAT'S THE DIFFERENCE BETWEEN WILDCATS AND DOMESTIC CATS?

Scientists can easily separate dogs from their wolf ancestors by genetic analysis, but the genetic differences between wildcats (*Felis silvestris*) and domestic cats (*F. catus*) are considered trivial and comparable to the differences seen among individuals within each of the two species. According to some authorities, the genetic evidence reveals that domestic cats are similar enough to both African and European wildcats to be called the same species; others disagree, maintaining the separation of the domestic cat from its wild ancestors. It is generally agreed on the basis of behavior, however, that domestic cats came from African wildcats. African wildcats are less fearful of people and less aggressive than their European counterparts, thus easier to tame. Today, African wildcats live near villages and show less fear of people than the far warier European wildcats.

Studies of red foxes reveal that selection for tameness and affection for people can turn a line of wild red foxes into doglike pets in a mere 20 years. Associated with this are other changes, including new coat color patterns, reproductive changes that resulted in their breeding all year around instead of once a year, and

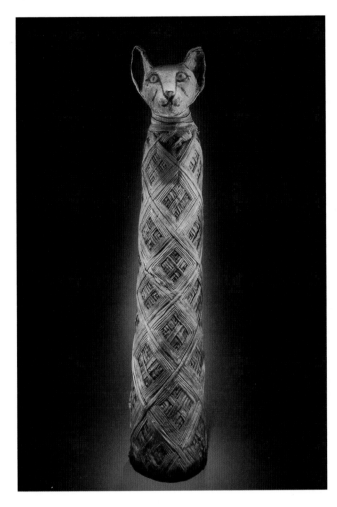

Ancient Egyptians mummified millions of cats as part of their devotions to the goddess Bastet, whom cats embodied. So common were these mummies that an English firm collected and pulverized an estimated 180,000 of them to sell as fertilizer in the late 1800s. (Photo by Chip Clarke)

higher levels of the neurotransmitter serotonin, which has a calming effect. Prozac and other drugs prescribed for depression in people increase the levels of serotonin as well. Veterinarians treat domestic cats with behavioral problems related to anxiety with similar drugs that increase serotonin.

Domestic cats similarly have diverse coat color patterns and may breed year around, whereas wildcats are believed to be seasonal breeders. Certain coat colors in domestic cats are associated with different temperaments. The most common coat colors today are blotched tabby, black, and orange, all colors correlated with calmer cats than the agouti or striped tabby, which have the ancestral coat colors. Domestic cats also tend to have proportionately smaller brains, jaws, and teeth, shorter legs, and longer digestive tracts than wildcats.

An interesting difference was recently discovered between domestic dogs and wolves. Both dogs and wolves equally socialized to people could find hidden food by following clues from a familiar person, such as touching or pointing to the food's location, but wolves were not as good as dogs. The two were also trained to solve a problem and then given an insoluble variation of the problem. Dogs turned and looked at the familiar person, as if to ask for direction; wolves did not, suggesting that dogs have evolved to communicate with people. People who have raised wildcats and individuals of other small wild cat species say that they are not afraid of people but simply indifferent to them. Wild cats in zoos ignore visitors, acting as if they aren't there at all. It would be interesting to conduct similar scientific studies to learn if domestic cats are any better at reading human signals or are more likely to look to people to solve a problem than tamed wildcats or other wild cats are.

WHAT ARE FERAL CATS AND HOW DO THEY LIVE?

Feral cats are free-ranging domestic cats that live outside of human control. In essence, these are domesticated cats gone wild again, something cats appear to do quite readily. Feral cats occupy diverse natural habitats, from the alleys of large urban areas, to farmlands, to remote oceanic islands where people no longer live or may never have been in permanent residence. They live in the cold temperatures of subantarctic islands and in the tropics. Also, their densities vary, for example, from about 1 per square kilometer in Australian grasslands to about 2,000 per square kilometer in the small fishing village of Ainoshima in Japan.

Discarded human food and other garbage form a large portion of feral cats' diet in urban areas where prey other than Norway and black rats, house mice (which are also attracted to human garbage), and various birds are scarce. Indeed, the line between feral and other domestic cats in urban areas is often indistinct, with some house cats left to fend more or less for themselves and some feral cats fed by people who don't take care of them in other ways.

On islands with no indigenous mammals, the primary prey of the introduced cats are other introduced species, including Norway and black rats, house mice, and European rabbits. On oceanic islands, seabirds that nest in large colonies are another source of food for feral cats. Elsewhere, feral cats, like the ancestral wildcats, are primarily predators of small mammals, from mice and voles to small (or young) rabbits and hares; compared with mammals, birds are a relatively small part of their diet in most places, and reptiles are taken only infrequently.

In some situations, feral cats live in well-organized social groups that are somewhat similar to, but more variable than, those of lions. For all animal species, variation in the distribution, predictability, and abundance of resources—food, water, shelter, nest sites, and mates—influences social organization. For most wild cats and many feral cat populations, food (prey) is widely dispersed and unpredictably found,

Domestic cats gone wild, feral cats often form social groups similar to those of lions, something their wildcat ancestors are never observed to do. Few places on Earth are without feral cats, which can reach astonishing densities where food discarded by people is abundant.

which results in their typically solitary life style. But where food is abundant and can be found in predictable patches, feral cats live in groups to take advantage of the largesse. These patches are generally associated with human activities, such as farms or garbage dumps.

The ability of feral cats to live in groups may, however, be an effect of domestication, because domestic cats have no doubt been bred to tolerate the proximity of companions, including people and other domestic animals. A study of African wildcats and feral cats sharing an area in Saudi Arabia revealed that feral cats formed a colony to use the clumped food resources of a dump, but the wildcats maintained their solitary existence.

WHY ARE FERAL CATS—AND PET CATS— SOMETIMES A MENACE?

Even with domestication, cats have lost none of their predatory instincts. Predation by feral cats has a significant effect on the numbers of rodents (rats and mice), lagomorphs (rabbits and hares), and birds in an area. Their strongest effect has been observed on islands but is felt on continents as well. The Global Invasive Species Specialist Group includes domestics cats on its list, 100 of the World's Worst Invasive Alien Species, noting that only habitat loss is a greater cause of species endangerment and extinction. Cats are believed responsible for, or at least involved in, the extinction of more bird species than anything else except habitat destruction.

One striking example shows how even a single cat's introduction to an island previously cat-free can wreak havoc. In 1894, David Lyall became the lighthouse keeper on a small, uninhabited island off New Zealand. He brought with him a cat. Among the cat's prey were unique flightless wrens, later named Stephens Island wrens. In typical cat fashion, the animal presented 17 of the birds to Lyall. These proved to be a species previously unknown to science—and remain the only specimens of this species anywhere. Within a year of Lyall and his cat's arrival, the wren was extinct. Unable to fly, the birds were easy pickings, and a single cat made short work of eating the species to extinction.

While not quite so dramatic, similar scenarios have played out elsewhere. On a subantarctic island chain called the Crozet Islands, cats extirpated 10 species of petrels, pelagic seabirds that nest on the ground on islands. In a fairly typical pattern, sealing ships accidentally brought rats and mice, and cats were then imported to control the rodents; rabbits were introduced to provide food for sailors, but they also provided food for cats, enabling the cat population to grow.

Bradford Keitt and his colleagues (2002) looked at the impact of feral cats on the black-vented shearwater, which is a bird endemic to the Pacific islands off the Baja California peninsula. About 95 percent of the world's population of this species lives on Natividad Island, along with a small number of cats. Estimating that each cat would eat about 45 shearwaters a month to fill its nutritional requirements, these scientists predicted that 20 cats could eliminate a population of 150,000 shearwaters, the size of the Natividad Island population in 1997, in less than 25 years. Fortunately, 25 cats—believed to be all the cats on the island—were removed after Keitt's study, before it was too late. Several subspecies endemic to these islands, however, had already been driven to extinction by feral cats.

Domestic cats arrived in North America with European colonists less than 500 years ago. Today, an estimated 70 million pet cats live in the United States, and another 40 million or so free-ranging, or feral, cats live in urban, suburban, and rural habitats. The number of small mammals and birds these cats eat is staggering: more than a billion are taken each year by rural cats alone! These cats do us a service in eating rats and mice that are pests, but they also take songbirds—by one estimate about 39 million birds are killed each year in Wisconsin alone—many of which are in decline because of habitat loss in North America as well as in Central and South America, where many of these birds winter. Cats are also implicated in the decline of least terns, piping plovers, and loggerhead shrikes.

A 2003 study in Florida revealed a host of species actually or potentially affected by pet and feral cats. For instance, the entire population of the endangered Florida Keys marsh rabbit is 100 to 300 individuals. Annually, some 53 percent of the deaths of this species are attributed to feral cats, and experts predict the species may be extinct within 20 or 30 years.

Predation is not the only problem that feral and pet domestic cats pose. Their potential to spread disease to wild animals as well as to people is also an issue (see *Do Cats Get Sick?*). Cats can spread rabies, feline panleukopenia (distemper), and FIV, feline immunodeficiency virus, which is similar to the human HIV. Pumas are known to have caught feline leukemia from domestic cats, and the critically endangered Florida panther subspecies may have been infected with distemper and FIV. In Israel, wildcats may be rare thanks to feline panleukopenia, which is spread by domestic cats there.

In 2003, scientists linked a parasite carried by domestic cats, *Toxoplasma gondii*, to many deaths of California sea otters from 1989 to 2001, when the population dropped by about 10 percent. The parasite is found in cat feces, and scientists speculate that runoff from the feces of feral cats and from flushable cat litter has transported the parasite to the ocean, exposing sea otters to it. Pregnant women who suffer toxoplasmosis as a result of infection with *T. gondii* may give birth to children with severe physical abnormalities.

Domestic cats have been implicated in transmitting bubonic plague to people in the United States, especially in Arizona, Colorado, and New Mexico, where suburbs are springing up near once-remote areas in which the disease is endemic in wild rodents. Domestic cats that roam freely in these areas are at increased risk for infection and, therefore, increase the risk for transmission to humans. Before 1977, domestic cats were not reported as sources of human plague infection; since then, domestic cats have been the source of infection in 15 human plague cases.

Feral cats also compete with native predators, such as raptors and other mammalian carnivores that depend on small mammals and birds for sustenance. One study estimates that cats are far more abundant than all the other medium-sized carnivores, or "mesocarnivores," combined in some rural Wisconsin areas. This must have an impact on the food available to foxes, raccoons, skunks, and other predators of this size.

In Europe, Asia, and Africa, domestic cats threaten their wildcat relative because they readily hybridize, and it is feared that domestic cat genes will swamp those of wildcats. In southern Africa, all wildcats living close to human settlements with

Both feral and pet domestic cats pose threats to populations of wild cats as well as to a host of other species. In Scotland, interbreeding between domestic cats and native Scottish wildcats (above) means that many, if not most, of the animals called wildcats there are actually hybrids. Similar hybrids appear even in remote areas of Africa and perhaps Asia as well.

domestic cats are hybrids, but hybrids have also been reported in more remote areas. Overall, the Cat Specialist Group believes this to be the greatest threat to the African wildcat. Scottish wildcats are also mostly hybrids, and Asiatic wildcat hybrids are reported from Pakistan and central Asia and probably occur elsewhere in Asia. This threat to wild cats may be greater than is known. Domestic cat fanciers have successfully mated domestic cats with rusty-spotted cats, Asian jungle cats, Geoffroy's cats, and other wild species to produce novelty breeds. Whether interbreeding occurs naturally where these wild species range near people is not known.

HOW MANY BREEDS OF DOMESTIC CATS EXIST AND WHERE DID THEY COME FROM?

For most of the history of domestic cats, there were no breeds, or at least none deliberately created. Unlike dogs, which people had long selectively bred to perform diverse useful tasks, from guarding sheep to hunting rabbits and pulling sleds, cats were left to be cats. Over time, differences emerged among the domestic cats predominant in various parts of world, thanks to the founder effect, in which the particular cats that reached an area were by chance somewhat genetically distinct from others in the general population. In addition, isolated populations adapted to local conditions through natural selection. From time to time, people may also have exerted some selection for novel or attractive fur color.

It wasn't until the middle of the nineteenth century in England that cat fanciers began to show cats and thus took an interest in selectively breeding them for particular qualities and importing exotic cats from afar. But even then, selection was for appearance and, later, temperament, not for functional attributes. The first-ever cat show was held in 1871 at the Crystal Palace in London, and now there are various associations of cat fanciers around the world. The Cat Fanciers Association, billed as the largest registry of pedigreed cats, recognizes 41 breeds; the American Cat Fanciers Association lists 46 breeds; and the International Cat Association counts 47. Desmond Morris (1996) describes some 80 breeds. The fact is, cat breeders work to create new breeds all the time.

In the mid-1800s, British cat fanciers divided domestic cats into two sorts: the British (also called American or European) and the Foreign. The British cats are cold-adapted robust, stocky animals with large heads, short ears, and thick fur; they resemble the Scottish wildcats, with which they interbreed. In contrast, the Foreign cats are a hot-climate form, with slender bodies, long legs, large ears, and short fur; they most resemble the African wildcat. In addition, domestic cats are divided into short-haired and long-haired types, with the Persian cat being an exemplar of the latter. The origins of the long-haired Persians are obscure, but this breed is believed to have arisen in Persia (modern Iran), just as long-haired Angoras originated in nearby Turkey. In any case, today's breeds are almost entirely the result of crossing British and Foreign cats and short-haired and long-haired cats, with additional selection for coat color, a fairly labile trait in all felids. A few breeds are the result of chance genetic mutations that breeders then selected for. For instance, the hairless Sphynx originated with a hairless kitten born in 1966, and LaPerms, which have curly fur, originated from a kitten born in 1982. A few others have their origins in recent crosses with wild cats. The Bengal, for instance, may have sprung from mating a domestic cat with a leopard cat.

Place-names attached to breeds may or may not reflect their geographic origins. Those that do are the names of many old breeds of cats that apparently emerged before cat fanciers began their experiments, including the Siamese, the Burmese, the tailless Manx from the Isle of Man in the Irish Sea, the Persian and Angora, as noted above, and perhaps the Abyssinian, from modern-day Ethiopia, a cat some believe to be descended from the first domestic cats of Egypt. In contrast, the Russian Blue may be called that only because the first of this sort to reach Britain came from Archangel, a Russian port. The Kashmir cat was created in North America, and Morris reports that "the name is based on the fact that this breed is close to the Himalayan breed and Kashmir is close to the Himalayas" (1996, 244). The Himalayan cat isn't from the Himalayas either; it is a cross between Persian and Siamese cats.

While these pedigreed cats owe their modern appearance primarily to human intervention, variations in coat color, in particular, appear to have ancient origins. The ancestral coloration of domestic cats is that of the African wildcat: striped or mackerel tabby. But black, blue, orange, white-spotted, and white cats are found worldwide, suggesting that this diversity was present when cats first began to leave Egypt.

WHY ARE CATS, ESPECIALLY BLACK CATS, ASSOCIATED WITH WITCHES?

In her 1998 book, called *The Cat and the Human Imagination*, Katherine M. Rogers wrote:

> Although it may seem obvious to us today that cats are peculiarly suited to supernatural roles, this attitude is largely a product of nineteenth- and twentieth-century sensibilities. The main reason cats were persecuted is the prosaic one that they were abundant and considered valueless. When animals were to be sacrificed as scapegoats to allay public guilt or anxiety, as placatory offerings to ensure a good harvest or a stable building, or as exciting special effects in spectacles and processions, the obvious course was to round up the local cats. (p. 46)

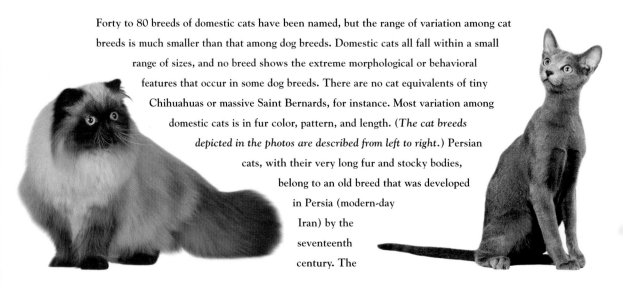

Forty to 80 breeds of domestic cats have been named, but the range of variation among cat breeds is much smaller than that among dog breeds. Domestic cats all fall within a small range of sizes, and no breed shows the extreme morphological or behavioral features that occur in some dog breeds. There are no cat equivalents of tiny Chihuahuas or massive Saint Bernards, for instance. Most variation among domestic cats is in fur color, pattern, and length. (*The cat breeds depicted in the photos are described from left to right.*) Persian cats, with their very long fur and stocky bodies, belong to an old breed that was developed in Persia (modern-day Iran) by the seventeenth century. The

Although we think of black cats and witches as natural companions, Rogers points out that it wasn't always so. Witches were believed to turn themselves into a variety animals when conducting evil deeds or to have an animal companion, "a familiar," that carried out their mischief. In England, for instance, the original witch's familiar was a hare, associated with Eostre, a pagan fertility goddess. Like cats, hares are nocturnal, elusive creatures; like hares, cats were pagan symbols of female fertility and sexuality. This last may help account for singling out cats for opprobrium. The early Catholic Church ruthlessly sought to eliminate vestiges of pagan belief, and it also condemned the female sexuality that cats symbolize. Furthermore, affection for a cat was in itself suspect. Cats' aloof indifference to people, their refusal to recognize people as masters, violated the natural order: God gave people dominion over animals, and thus cats, with their secretive ways, must be on the side of Satan.

From the twelfth through the fourteenth century, the Church accused heretics of worshipping Satan in the guise of a large black cat, and, later, witches were accused of flying to their nocturnal meetings on the backs of large black cats. But during the heyday of European witch persecution, in the 1500s and 1600s, legal records show that a woman might be accused of witchcraft if she associated with cats of any color, as well as with dogs, mice, toads, lambs, rabbits, and polecats. Rogers writes: "Any animal would do if it were small—because of the physical intimacy between witch and familiar—and cheap. Accused witches were usually too poor to own highbred pets, and it was considerably safer to accuse a poor woman's cat of being an agent of Satan than to accuse the squire's prize greyhound" (1998, 52). Cats became typecast in the role of witch's companion only when witchcraft became the

breed known as Russian Blue is short-haired and tabby, with its markings masked by dense bluish-gray fur; its angular body reveals the recent introduction of Siamese cat genes into the breed. The first domestic cats were tabby, but only lightly striped and spotted. The blotched tabby, with its strong markings and short hair, displays a common fur configuration of domestic cats today. This form probably first appeared in Britain about 400 years ago and spread from there across the globe with the growth of the British Empire. The freakish hairless Sphynx is a product of the twentieth century. (Photographs by Richard Katris, Chanan Photography) stuff of literature, and writers found cats more attractive as familiars than other small animals. By the 1800s, when belief in witchcraft was largely a thing of the past, witches and their cats began to acquire an exotic allure.

Before this shift occurred, attitudes toward cats remained mostly negative. For centuries, cats were the pets of the poor: compared with dogs, they required no extra feeding and demanded no special care. Although both animals were useful to people, dogs were admired for their subservience to humans, while cats were disliked for their lack of it. Depicted in paintings, cats represented greed and a threat to domesticity. Cats were casually persecuted and tortured with shocking indifference, both officially (they were burned alive inside an effigy of the pope during Elizabeth I's coronation) and at the hands of small boys who might fling them from windows or tie two together by the tails in order to watch them fight. Even influential naturalists reviled cats. In 1607, Edward Topsell's *The History of Four-footed Beasts and Serpents and Insects* painted cats as dangerous beasts with venomous teeth and poisonous flesh that caused illness and joined in branding cats as familiars of witches (cited in Rogers 1998).

The Romantic movement, which began in the 1700s, changed all that. Romantics glorified wild, untamed nature, and suddenly the cat became the dog's superior. Dogs were denigrated for their servility; cats were celebrated for their insistence on freedom. Cats "could not be fully appreciated as companions until their wildness was perceived as attractive rather than obstinately perverse," says Rogers (1998, 189). A greater appreciation for cats finally led to the creation of breeding societies in the late 1800s, when, for the first time, cats were valued enough to be shown by the wealthy and bred for desirable and attractive traits.

WHAT ARE SOME SUPERSTITIONS ABOUT CATS?

While purported witches did not confine themselves to black cats, black cats did epitomize feline evil, their color alone representing the forces of darkness. Black cats have signified bad luck for centuries, and even today, superstitious Americans consider a black cat crossing their path as bad luck. In Britain, however, this is a sign of good luck, based on the idea that evil has passed you by. And in many other contexts, such as on ships and backstage at a theater, cats are good luck; a cat on stage, however, is bad luck. Superstitions arise when some event is coincidentally associated with another one several times, and people mistake the coincidence for cause and effect. Given the abundance of cats, it's not surprising that both good fortune and misfortune might often befall people while in the presence of cats.

Other superstitions about cats relate to their effect on human health. Cats were once believed to cause rheumatism and tuberculosis. People still believe that cats suffocate babies or suck their breath. These beliefs may have some basis in many people's allergic reaction to cats, or, more specifically, to a protein called fel d 1, found in cats' saliva and sebaceous glands. When a cat grooms itself, its hair and skin become coated with this protein, which is then shed as dander and inhaled by anyone in that environment. The airborne protein triggers an immune response in susceptible people. The allergic reaction varies from irritating itchy eyes and runny nose to life-threatening asthma attacks. Cat allergies are common around the world; in the United States, 5 to 10 percent of the population experiences an allergic reaction to cats, and a major study found that 30 percent of laboratory staff working with cats develop allergies. A recent study, however, suggests that not all cats are equally allergenic. Scientists studied 312 patients who reported severe allergy symptoms, moderate symptoms, mild symptoms, or no symptoms. Subjects with dark-colored cats were two to four times more likely to report severe or moderate symptoms than those with light-colored or no cats. There was no difference in the severity of the symptoms between those with light cats and those with no cats. The scientists speculate that black cats may produce more of the fel d 1 allergen than others. Thus, for people susceptible to allergies, black cats may indeed be unlucky.

Finally, many superstitions about cats relate to the idea that cats can predict the weather. For instance, to some, a cat washing its face predicts good weather or is a sign of rain, and so is a cat sneezing or scratching a table leg. When a cat's routine behavior predicts both good and bad weather (or good and back luck), it is bound to be correct half the time, a good basis for the development of superstitions. Some speculate, however, that cats' sensitivity to vibrations and to sounds beyond our hearing enables them to detect coming storms or earthquakes. This hypothesis is attractive until you consider the number of other animals believed to be able to predict the weather. For instance, a whistling parrot, a hooting owl, a quacking duck, a sitting cow, and a noisy sparrow are all thought to predict rain.

DO CATS ATTACK AND KILL PEOPLE?

People who admire large cats and worry about their future don't like to think of them as enemies. But big cats do kill people. Resolving this conflict and devising a coexistence recipe are essential if large cats are to survive in our human-dominated world. On the other hand, it is probably good for us to lay our hubris aside and think of ourselves as just another meal from time to time, because it helps us to understand our world more comprehensively as well as our place in it.

Tigers, lions, leopards, pumas, and jaguars sometimes kill people. Cheetahs, snow leopards, and clouded leopards have never been reported to do so. Cases of tigers killing humans come from almost all parts of their range in Asia. (The term *"man"-killer* is traditional, although perhaps even more women than men are killed by large cats. Also, in some historical traditions a man-killer was not recorded and the big cat was not labeled as a man-eater until a number of victims—three or more in some places—had been killed.)

Cases of lions killing people come from nearly their entire range in Africa. The Asian lion was not known for man killing, but recently several cases have been reported in the Indian state of Gujarat, where the last Asian lions live. Leopards are known to kill people throughout most of their range in Africa and Asia but, under most circumstances, less often historically than lions and tigers have killed. Recently, with the tiger in decline in many places it formerly lived, leopard numbers have increased, and so have human deaths attributed to leopards in those areas. This has been particularly troublesome in Nepal and northern India.

Biologist Paul Beier searched all the records he could find from 1890 to 1990 for incidents of puma attacks on people, and he was able to document 9 fatal attacks and 44 nonfatal attacks resulting in 10 human deaths and 48 nonfatal injuries (two victims were involved in each of five different attacks). Between 1991 and 2003, more puma attacks on people occurred, several of them fatal. There are very few records of jaguars attacking people. Kevin Seymore concluded in a major review that "the jaguar is least likely of any of the pantherines to attack man and is virtually undocumented as a maneater" (1989, 5). Biologists Rafael Hoogesteijn and Edgardo Mondolfi have extensive experience with this large cat and conclude much the same: "Although there are numerous cases of attacks on man, the overwhelming majority are due to hunting accidents in which the feline was wounded or cornered by dogs" (1993, 88).

The most comprehensive account of man killing by tigers and leopards is by environmental historian Peter Boomgaard. In *Frontiers of Fear: Tigers and People in the Malay World, 1600–1950* (2001), he compares man killing by these two large cats in India and the "Malay world," which, for his purposes, includes peninsular Malaysia, Sumatra, Java, and Bali. Tigers but not leopards have occurred on the islands of Sumatra and Bali, but both have occurred on Java and peninsular Malaysia and in India. His sources were Dutch and British colonial records. His findings for the years 1882 to 1904 are astonishing. In India, an annual average of 889 deaths were attributed to tigers, and 317 to leopards, for a total annual average of 1,206. In Sumatra and Java, respectively, 58 and 51 annual deaths occurred during this period. Using 1875 as a benchmark, Boomgaard reported that for every million people, there were 66 deaths by tigers and leopards combined in India, 54.5 deaths

The determined approach of a huge lion would terrify anyone who stood in its path. Lions, tigers, and a few other big cats do sometimes kill people and have done so throughout our shared history. Man-eating cats, which specialize in preying on people, are rare, however. Most of the tragic human deaths caused by big cats result from chance encounters that provoke a cat to attack.

by tigers in Sumatra, and 6.1 deaths by leopards and tigers combined in Java. Annually from 1860 to 1900, about 5,000 tigers and leopards were killed in India; 400 tigers and clouded leopards in Sumatra; and between 1,400 and 1,496 tigers and leopards in Java. Today, by official estimates, fewer than 3,000 wild tigers live in India, fewer than 500 tigers in Sumatra, and the tiger is now extinct on Java and Bali.

Although the British began their settlement of the island of Singapore in 1819, they reported no problems from tigers until they began clearing the island for plantations of peppers and other crops. By 1850, it was purported that one laborer a day was killed by tigers. There are no tigers on Singapore today. The British began to expand their colonization of the Malay Peninsula later, and records from there are not very useful. Livestock are reportedly killed each year by tigers in Malaysia, but human deaths rarely have been reported. In the last decade, people have been killed by tigers in Nepal, India, Bangladesh, Sumatra, and the Russian Far East.

WHY DO CATS ATTACK PEOPLE?

The more we know about the behavioral ecology of large cats and why some become man-killers, the more remarkable it seems that attacks do not occur more often than they do, especially given the large number of people who are sharing landscapes with large cats. But there are more ideas about why some large cats kill people than why others do not.

We don't know how many attacks on people occur because the cat has the furious form of rabies, although at least some are likely to stem from this. Large cats may kill people simply because their prey have been depleted and they are hungry. Or, the cat may be hungry because it is incapacitated or compromised by an injury and is unable to kill its normal prey. Or, the cat may be socially subordinate and pushed into marginal hunting areas by dominant cats. Or, these subordinate cats may hunt during the day, increasing the odds of meeting a person out and about. Young cats in the process of learning how to hunt often go after prey that an experienced cat would not. For example, cheetahs with no actual hunting experience stalk and rush at objects as diverse as policemen on motorbikes, Grevy's zebras, and yellow school buses at the National Zoo.

In other cases, a large cat may hunt livestock or dogs around people's dwellings or farmsteads and have a fateful encounter with someone who steps outside. John once investigated an incident in India in which a tiger killed a woman who had gone out at night to relieve herself. The tiger was probably after goats tied under her porch, and it was just bad luck that she appeared at the same time. The tiger had dispersed from the forest and was essentially trapped among some villages and shrimp ponds in the Sundarbans, the mangrove forest at the mouth of the Ganges in India and Bangladesh.

People may unknowingly present the right stimuli at the right time in the right place for a hungry cat to be motivated to kill. A kneeling or squatting person looks more like a deer than a standing one, for instance. A hunting, hungry puma sees a child bounding ahead of his parents down a mountain trail at the edge of the forest in the late afternoon, and this sets off the cat's predatory sequence. A runner alone on a forest trail may attract a hunting puma's attention. Or a hiker who kneels down to tie a shoelace or a rubber worker, honey gatherer, or grass cutter who bends down while going about his work triggers a tiger attack. These people may be providing just the right set of stimuli to set off the approach-and-seize sequence in a watching cat.

In some circumstances it appears that eating people is learned. One tigress and her cubs began to feed on human remains left on the river banks of the Holy Ganges after partial cremations, which are part of Hindu death ceremonies. Soon, she and her cubs began to kill people.

By examining the teeth and jaws of lions in museum collections, biologist Bruce Patterson and his associates recently tested the infirmity theory as a general explanation for lions attacking people (Patterson et al. 2003). About 40 percent of the lions they examined had damaged teeth, and almost every old lion had broken teeth, including some with exposed roots. They compared their sample of museum specimens with a sample of the skulls of problem lions that had been killed by authorities in Kenya. The problem lions had many fewer incidents of dental damage than the lions in the museum sample. These were mostly younger animals, apparently in excellent health, taken where they had come into conflict with people on farms that bordered parks. These biologists conclude that the infirmity theory may explain some lion attacks, but most are linked to lions killing livestock, people, or both when prey is depleted. These younger problem lions may have been social subordinates and also may still have been learning the fine points of what and how to hunt. John and his associates came to a similar conclusion after a recent rash of man-killing incidents by tigers in Nepal.

Man killing is a fact of life in some areas, including the Sundarbans. There are no permanent settlements in this mangrove forest, but tens of thousands of people enter the forest legally and illegally to log, to collect firewood, honey, and other forest products, and to fish. Here, people are often by themselves for long periods of time, and some simply disappear. But some are known to have been killed by tigers, for as many as 100 deaths from tiger attacks are reported each year from this region. Local people believe that tigers actually hunt humans who come into the forest. It may be that some tigers learn to hunt people in this situation and continue to do so because there is nothing to teach the tiger not to. In most places, a tiger turned man-killer is found by the local authorities and captured and put into captivity or killed.

In some areas, after subtle changes, man killing increases where it has not been a threat previously. In the Gir Forest, in the Indian state of Gujarat, the last Asian lions have lived for years in a kind of mutualistic relationship with local people living in the forest. The Maldharis are people who live in small compounds called *nesses* scattered through this tropical dry forest. These compounds include a few stick-and-thatch houses and stockades of acacia thorn bushes, where they keep their domestic water buffalo and cattle for the night. They live a nomadic life style, burning these compounds every few months and constructing new ones. The primary livelihood of these people is selling the cooking ghee obtained from heating buffalo's milk butter. Lions, however, enter the stockades at night and kill and drag away both buffalo and cattle. The Maldharis graze their buffalo in the forest, and, on a fairly regular basis, lions kill buffalo there as well. In all, about 75 percent of the lions' diet was Maldhari cattle and water buffalo in the late 1960s.

The Maldharis are devout Hindus and cannot handle a cow or buffalo carcass for religious reasons, so they do not try to retrieve the hide or meat from lion-killed buffalo themselves. Instead, they drive the lions away by throwing rocks at them and then call in a group of hide-and-meat collectors, the *harijans*, or untouchables. The harijans, also keeping the lions at bay with well-thrown rocks, skin the carcass, take some meat for themselves, then leave the remains for the lions (in this case, people are kleptoparasites of lions). If the lions back off too far, the vultures will clean up the skinless carcass in less than an hour.

This is a real training session for lions. John has watched a Gir Forest guard armed only with a walking stick lead a buffalo calf down a dusty forest road with a whole pride of lions following along behind. Lions were whacked with a rock if they did not stand back and were rewarded with the calf or its carcass after the guard left. This has been a way of life for both lions and people generation after generation, and previously in that system there were few reported incidents of lions killing people. However, the system has not withstood the changing times. There are more lions now (up from fewer than 200 to about 300) after a conservation effort reduced the number of domestic livestock grazing in the forest, making more fodder available to the lion's natural ungulate prey. As a result, the natural prey have increased, and so have the lions. Furthermore, the lions are no longer trained to keep a distance by people pelting them with rocks to protect their livestock. There are now more incidents of lions attacking the people who still sneak into the forest to graze their livestock, and increased incidents also occurred when the forest was recently overrun by herders and their livestock during a drought.

HOW CAN SOMEONE AVOID A LARGE CAT ATTACK?

Our own antipredator behavior is the key to avoiding a big cat attack. Healthy respect is essential. Traveling in groups is useful. Remember the cat's hunting sequence: a readiness to hunt, search, find, stalk, seize, kill, consume. If you surprise a cat that is not hunting, which is unlikely, you should do the same things that you would do if you somehow saw a cat stalking you. Make yourself as unlike prey and be as aggressive as possible. Face the cat and stand tall. Don't turn and run away. Yell at it. Wave your hands; make yourself look bigger than you are. Throw stuff at it, but be sure you know where the cat is and what it is doing before you take your eyes off it to pick up a branch or stone. If you have a child, pick him or her up in your arms.

Paul Beier's analysis of the victims of puma attacks provides a useful framework for avoiding an attack. Thirty-seven of 58 victims were children; of these, 35 percent were alone; 43 percent were in groups of other children; 22 percent were accompanied by adults. Eleven of 17 adults attacked were alone. Except for one child and one adult who died of rabies, all fatalities involved children unaccompanied by adults. Most victims did not see the puma before they were seized. In some near-attacks, the puma was shot as it approached. Other near-attacks were thwarted when the person shouted, swung a stick, waved arms, and threw rocks. (John once did all of these things to deter a puma that was advancing on him.)

Behavior that failed to avert attacks included just watching or running away. One victim initially stood her ground, but when the two pumas continued to advance, she scrambled up an embankment and climbed a tree. The pumas continued to advance until she kicked one with her foot and struck the other with a stick. Advice to play dead is often given: Don't follow it. In one case, an eight-year-old followed that advice when it was shouted by his father. The puma started dragging the child off until his mother came up and screamed at it. As puma biologists E. Lee Fitzhugh and David Fjelline reported: "We know of no instance anywhere in which an attack, once contact was made, was stopped when the victim feigned death" (1997, 26).

WHO ARE SOME FAMOUS MAN-EATERS?

Man-eaters become famous because hunters of man-eaters write their story or, rather, tell of the intensity of the relationship between themselves and the animal, from the human point of view. These stories, of course, always end in the death of the man-eater, not the other way around. By far, the most famous men to describe their experiences in ways that are both entertaining and informative are Jim Corbett, a railway executive who hunted man-eating tigers mostly in northern India in the years before the Second World War, and Lt. Colonel J. H. Patterson, who killed two notorious lions that ate laborers building the Mombasa to Nairobi railroad in 1898. Patterson's lions were dubbed—and his book named—*The Man-eaters of Tsavo*. In his book, *Man-eaters of Kumaon*, Corbett named many of the man-eating tigers and leopards he killed: the Mohan Man-eater, the Kanda Man-eater, the Pipal Pani Tiger, the Thak Man-eater, and the most famous of all, the Champawat Man-eater. The Champawat Man-eater was a tigress said to have "been driven out by a body of armed Nepalese after she had killed two hundred human beings" (1946, 4). She continued to kill people in India. Corbett shot her and determined that she had suffered an old gunshot wound to her mouth that had broken her canines. There was much speculation about the Tsavo man-eaters. The skulls, now in the collection of the Field Museum of Natural History in Chicago, were recently examined by biologists Bruce Patterson and Ellis Neiburger. Both were male; one had a broken lower canine and had developed a severe root abscess, probably preventing it from making killing bites on prey it would normally have hunted.

HOW DID OUR DEFENSELESS ANCESTORS GET ALONG WITH BIG CATS?

Our antipredator behavior is augmented and modified by culture, including the tools we have at our disposal. Humans have been dealing with the threat of large predatory cats for as long as we have been around, and we do so in many ways. Many anthropologists now believe that many of the large mammal remains found with early humans at fossil sites were scavenged from other carnivores. Even today, human kleptoparasitism of large cat kills is standard in some groups of people. In Nepal, trained elephants are taken to the forest each day to feed. If the remains of an ungulate killed by a leopard or tiger are encountered, the elephant keepers appropriate it. If the tiger is present, they use their elephants to push the tiger away and then take the kill. The Hadza people of northern Tanzania take kills away from leopards, lions, and hyenas. They find the kills by watching vultures, and if the predator does not flee when they approach, they shoot arrows at it. Lions do not always retreat, so they become hunting targets themselves.

According to Boomgaard (2001), on the Malay Peninsula and Sumatra, indigenous people used both spears and blowpipes with poison-tipped darts to hunt tigers. Other groups used snares to capture and kill both tigers and leopards. Others poisoned tigers by treating fresh tiger kills with *walikambing,* which is derived from the plant *Sarcolobus spanoghei.* Boomgaard reports that these people nearly always had an active tiger trap of some kind near their villages, usually box traps, pits, trip logs with a spear imbedded in them, dead falls, and, more recently, trip-wire guns.

The people who seem most affected by large cats are not those who have lived with them for eons, but rather newly arrived immigrants to a region where big cats live but the immigrants have no cultural experience dealing with them. This was the situation faced in the Asian colonial era, when plantation owners used large numbers of people to work on newly established plantations and to construct public works projects, such as the railroad mentioned in *Who Are Some Famous Man-eaters?* Clearing forested areas for new plantations attracted ungulate prey, and so large cats concentrated their hunting there, creating the perfect environment for confrontations between big cats and people. Tigers and leopards, in just these circumstances, killed many people. Other peoples were displaced from their traditional land by the new plantations. Even if they did not move far, they suddenly found themselves facing large predators for the first time. Without any cultural knowledge about how to prevent large cat attacks, many people died. We were struck by an account of nineteenth-century east Javan farmers wearing large baskets in an attempt to protect themselves from tiger attacks when they went into their fields to plant rice.

The issue of inexperienced people in tiger, lion, and puma country is with us today in full force. In the western United States, people are building suburbs right up next to the mountains in some areas. This is winter range for deer and elk, and people enjoy seeing these animals in their backyards. But the areas where they have built are also puma hunting sites, which has led to increased human encounters with these cats. In Africa, ecotourist camps have been established in lion country, putting people right next to lions, which has resulted in increased attacks on humans.

WHAT ROLES HAVE LIONS PLAYED IN DIFFERENT CULTURES?

No other group of animals exerts a more powerful influence on people or compels more human interest and fascination than cats do. Our relationship with cats, whether predatory tiger or purring house cat, is ancient and universal.

According to Greek legend, the sun god Apollo created the lion; and his twin sister, Artemis, who was identified with the moon, created a miniature copy of the lion—the cat—to ridicule Apollo. This legend shows that people have long recognized the similarity of lions and small cats, despite their enormous disparity in size, and it suggests that part of the appeal of domestic cats is that they symbolize mastering the awesome big cats. The legend also goes on to encapsulate cats as symbols of sun and moon, good and evil, life-giver and life-taker.

Lions were already in possession of the African plains when our ancestors first emerged from the forest to embark on a new way of life. To these poorly armed and virtually defenseless protohumans, lions must have loomed large—as predators, as competitors, and even as teachers in the ways of hunting. And throughout human evolution, people have encountered lions the world over, before their range became limited to what it is today (see *What Cat Has the Widest Distribution?*).

Lions occupy a dominant role paralleled only by the domestic cat in the myths, symbols, legends, literature, and art of peoples throughout the world. Portraits of lions appear in Europe's Paleolithic cave paintings and engravings. The earliest portrait of the Egyptian goddess Bastet, dated 3000 BCE, shows her as lion-headed, although later she was more often revealed as a cat. Since that time, lions have been depicted in every media and appear on objects from ancient coins and jewelry to modern cars and baseball caps. Wherever they are used, lions symbolize strength and power. They are often associated with the sun. Bastet as a lion-headed goddess was described as the "flaming eye of the sun" but was considered more benign than

Cats figure prominently in the cultures of the people around the world, often symbolizing the forces of good. In Bali, Indonesia, a mythical lion called Barong, depicted in this mask, represents the sun, light, right, and deliverance from disease. The good Barong is said to have helped the people to thwart an evil witch—representing darkness, illness, and death—in her plans to destroy them.

her fierce lion-headed twin, Sekhmet, who was associated with war as well as the sun. In Mesopotamia, the lion was the symbol of Ishtar, the goddess of love and war. Much later, in medieval Europe, the lion appeared as a symbol of Christ.

Coins from the Mogul Empire in India depict lions and the sun together. In Africa, a staff carved with a lion figure often signified a chief's power, and the belief that eating lion meat, especially a lion's heart, imbued people with great courage was widespread. Also widespread in Africa were beliefs about chiefs reincarnated as lions and the souls of the dead entering the bodies of lions. Similarly, the ability to kill lions was a testament to a hunter's prowess. Among the Masai and other African groups, killing a lion is a rite of passage for young men, although it is now illegal to do so. Heracles, Samson, and kings throughout history earned powerful reputations through hunting lions; interestingly, the hunting of these big cats was often forbidden to anyone but rulers. Many individuals—from Europe's Richard Lion-Hearted to Ethiopia's Haile Selassie, dubbed Lion of Judea—adopted the lion as a symbol of power. Lions also emblazon European flags and coats of arms, and in the Netherlands, the lion is the national symbol. Today, lions continue to represent power—as mascots for football and other sporting teams, as corporate logos, and in advertising.

Lions are sacred to Buddhists as defenders of the law and as protectors of sacred buildings. In the latter capacity countless stone lions guard castles, churches, bridges, and public buildings throughout the world. Perhaps the notion of lions as defenders of the law is extended in the theme of lions as magnanimous beings. This theme recurs throughout literature, from the famous story of Androcles and the lion to the modern children's story *The Lion, the Witch, and the Wardrobe* and the animated film *The Lion King*.

WHAT ROLES HAVE TIGERS PLAYED IN DIFFERENT CULTURES?

Tigers play a powerful symbolic role in Asia similar to that of lions in Africa and western Europe. Koreans dub the tiger King of Beasts, and throughout Asia, tiger motifs connote the power of kings. In China, the markings on a tiger's forehead are interpreted as the Chinese character for "king"; tigers and dragons together symbolize the two great forces of nature—yin and yang. As with lions in Africa, the hunting of tigers in Asia was a sport reserved for potentates. Tigers were often kept in oriental courts and used as executioners, as lions were in Rome. Chinese emperors employed tigers in boar and deer hunts.

Tigers play a role in the religious beliefs of Asia. In Hinduism, Shiva is both destroyer and reproducer; as destroyer he is pictured wearing a tiger skin and riding a tiger. His consort is Parva the Beautiful, who, in her dark side, appears as Durga the Terrible riding a tiger. A disciple of Buddha also rides a tiger to demonstrate his supernatural powers and ability to overcome evil.

Asians are profoundly influenced by the tiger and its danger to people, and their belief systems reflect this. Those who live in the tiger's domain view tigers with great respect as well as with fear. Many forest tribes in India have deified tigers and erected temples and other shrines for tiger worship. In Thailand and Malaysia, forest-living aborigines believe tigers to be the avengers of their Supreme Being; tigers will kill only those who violate tribal law. In Sumatra, where Islam was introduced 500 years ago, the tiger is believed to punish sinners on behalf of Allah. Throughout the villages of Southeast Asia, shamans (priests) and magicians adopt the guise of tigers to present a powerful, fearsome image.

Like lions in Africa, tigers in Asia are commonly associated with soul transfer and reincarnation, and legends of were-tigers (people who turn themselves into a tiger at will) are widespread. In some Asian cultures, groups of were-tigers are believed to live, in their human form, in complex village societies.

In modern Asia, tigers figure prominently in advertising (as they do in North America and Europe), and tiger brands of products from soap to matches to beer abound.

Being an Asian animal, the tiger appeared later than the lion in European culture. But once this species became known, through travelers' tales and through animals brought to Europe, the tiger's sinuous beauty and awesome power quickly fired the imagination of writers and artists. Shakespeare often used tiger images, for example, when Romeo expresses his desire for Juliet: "The time and my intents are savage-wild, / More fierce and more inexorable far / Than empty tigers or the roaring sea." In 1936, Winston Churchill drew on the Chinese proverb "He who rides a tiger is afraid to dismount" to warn of the impending dangers in Europe: "Dictators ride to and fro upon tigers which they dare not dismount. And the tigers are getting hungry." In the 1980s and 1990s, when the economies boomed in several Asian nations, including Thailand, Indonesia, South Korea, Malaysia, Singapore, Taiwan, and Hong Kong, these nations were nicknamed the "Asian tigers"; inevitably, when their economic growth stalled in the late 1990s, headline writers referred to these "tigers" as being tamed or even skinned.

WHAT ROLES HAVE JAGUARS AND PUMAS PLAYED IN DIFFERENT CULTURES?

In the cultures of the Americas, jaguars and pumas occupy symbolic roles similar to those of lions and tigers in other parts of the world. From Mexico to the Andes, art, religion, and culture were historically dominated by jaguar images. Throughout Maya history, the jaguar symbolized the night sun of the underworld, which personified fear, night terrors, and death; jaguar motifs thus figured prominently on Maya funerary vessels. The Tucano Indians of the Amazon believed the sun created the jaguar as his earthly representative. The Olmecs of Mexico and many other peoples deified jaguars and built massive monuments devoted to their worship.

Everywhere, jaguars were associated with power and religion. Rulers in ancient Guatemala were given the title "jaguar." Shamans were believed by some Indians to turn into jaguars after their death, and it was widely believed in Mexico, in an echo of the were-tiger legends of Asia, that some people could take on the shape of jaguars. The Arawak Indians in northeastern South America today still perform man-jaguar transformation rituals to give a shaman the power to bring good or evil as the situation demands.

In the South American Andes and in North America, pumas rather than jaguars have featured as both feared and respected deities as well as cultural symbols. The ancient Peruvian city of Cuzco was laid out in the shape of a puma. Just as the

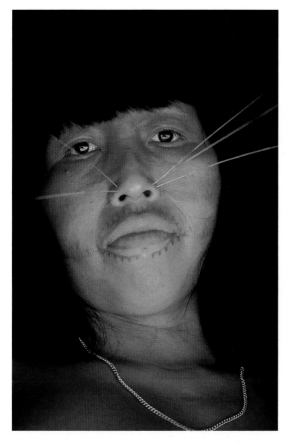

Hunters and gatherers native to the Amazon rain forest of Peru, the Mayoruna people revere the jaguar, the largest, most powerful cat in the Americas. The blue tattoo marks across this woman's face mimic a jaguar's grin, and the palm splinters embedded in her nose represent a jaguar's whiskers. These people believe that looking like a jaguar gives them the strength and prowess of the big cat.

jaguar represented the Maya underworld, so a widespread North American Indian legend describes the puma in the form of an underwater panther ruling the watery underworld and controlling storms. The Cochiti Indians of present-day New Mexico carved a pair of life-sized puma statues into the bedrock atop a mesa, creating a puma shrine that a few native Americans still visit today.

For European Americans, jaguars and pumas, as the largest cats in the Americas, took on the mystique of lions and tigers. American big-game hunters, for instance, looked to jaguars and pumas for the ultimate trophy. Jaguars and pumas have also come to represent fast cars and football teams.

WHAT ABOUT THE ROLES OF OTHER WILD CATS IN HUMAN CULTURES?

Everywhere that cats are found, people seem to have incorporated them symbolically into their culture. Leopards are common totemic animals in Africa, and were-leopard legends are known from Africa and India. In the Neolithic town of

A traditional Kikuyu dancer in Kenya drapes himself in a cheetah skin. Capes and coats made of cat fur form the ceremonial apparel of many traditional people. Although such uses usually have had little impact on wild cat numbers, as cats decline for other reasons, these practices may have to be curtailed, contributing to the erosion of cultural diversity.

Çatalhüyük, an archaeological site in present-day Turkey, evidence exists of leopard worship; also, there and at other sites, leopards have been associated with fertility goddesses. Leopards are common elements in European heraldry, and some experts believe that the lions that symbolize England were originally leopards.

In Norse mythology, Freyja, the goddess of love and beauty, is associated with the lynx and is often depicted riding one. The golden cat is sacred to some central African tribes, and the clouded leopard is to some Southeast Asians. In various societies, cheetahs and caracals were once trained to stalk and capture prey. In parts of Europe and the Middle East, cheetahs once rivaled dogs as the favorite hunting companions of the rich.

Cats everywhere, especially big cats, evoke powerful emotions in humans—fear, awe, and longing. The image of beautiful, majestic big cats living as mighty predators is absolutely compelling, representing all that is powerful, wild, free—so compelling, in fact, that some big cats live only in our imagination: the extinct Bali tiger, for example, still exists in the minds of many Balinese; and, on a lighter note, erroneous reports of a loose tiger or leopard regularly electrify rural communities in the United States. Sadly, without human action to conserve great cats, the awe these magnificent beasts inspire, and have inspired throughout history, will exist only in our collective memory.

WHAT PRACTICAL USES DO PEOPLE HAVE FOR CATS?

Although the many ways that cats figure in the world's cultures are extremely valuable to us, people put various cats to practical uses as well. The domestic cat is well known for its role as a rodent-killer, but little known is its performance of this function in some strange places. Desmond Morris (1996) described "refrigerator cats," also called Eskimo cats, a breed that was developed in the nineteenth century to thrive in the icy temperatures of commercial refrigeration plants and combat the rats that had already adapted to the frigid temperatures. Cats are still welcomed in libraries, where mice are apt to nibble on books. An organization called the Library Cat Society, founded in 1987 and based in Minnesota, keeps track of library cats around the world. Its Web site reports that, as of early 2003, there were 150 library cats in the United States, including 22 actual cats that reside permanently in libraries, and many others—15 statues, 2 virtual library cats, 2 stuffed lions, 1 stuffed Siberian tiger, 1 stuffed cheetah, and 1 ghost cat—that only symbolically threaten mice.

Cat Cuisine

In *The Voyage of the Beagle*, Charles Darwin (1845) described eating puma in Patagonia and mentions reports on the taste of other cats as well:

> At supper, from something which was said, I was suddenly struck with horror at thinking that I was eating one of the favourite dishes of the country namely, a half-formed calf, long before its proper time of birth. It turned out to be Puma; the meat is very white and remarkably like veal in taste. Dr. Shaw was laughed at for stating that "the flesh of the Lion is in great esteem having no small affinity with veal, both in colour, taste, and flavour." Such certainly is the case with the Puma. The Gauchos differ in their opinion, whether the Jaguar is good eating, but are unanimous in saying that cat is excellent.

While nowhere do cats form a dietary staple, some people do eat cats at times. *The Oxford Companion to Food* reports that lion meat is enjoyed in Africa and Asia (though in Asia this can include only India) (Davidson 1999). In Africa, where the growing trade in bushmeat for human consumption is threatening forest-living species from gorillas to porcupines, golden cats and leopards are sought after in some areas or killed opportunistically when hunters encounter them. In China, the imaginative, some might say bizarre, cuisine of Canton features wild and domestic cats (as well as just about every other animal there is). One dish described as "dragon and tiger head" is actually a stew of snake and domestic cat. In the wake of the outbreak of SARS in 2002 and 2003, which was linked to the consumption of masked palm civets, the government of China is trying to close down its exotic wild meat markets, but it is likely that cat eating will continue there to some extent.

In Asia, however, only a blurry line separates food from medicine, and cats, especially tigers, have been considered powerful traditional medicine. Nearly every part of a tiger, from its eyeballs to its tail, may be eaten or carried as a talisman to treat or prevent various maladies. A soup that includes tiger penis is served to promote sexual vigor—a traditional version of Viagra. Powdered tiger bone is used to treat arthritis and rheumatism. Trade in tiger parts and products is illegal throughout the world, but poaching continues to serve the demand for these products. Moreover, with tigers being extremely rare, leopard and other wild cat bones are being substituted for tiger bones.

People in Cat's Clothing

Although medicinal use of cats has threatened primarily tigers, killing cats for their beautiful fur drove down the numbers of many species in the last century. In the past, cat fur was fashioned into capes and coats, worn largely for ceremonial purposes in various cultures. In the twentieth century, however, spotted cat fur coats became fashion statements in the United States and, to a lesser extent, in Europe. Various reports estimate that the pelts of more than 10,000 leopards, 15,000 jaguars, 3,000 to 5,000 cheetahs, and 200,000 "ocelots" (in this context, a generic term for this and similar species, such as margays) were imported into the United States and Europe during the late 1960s. Although poaching remains a problem, most trade in cat fur is now illegal. Moreover, largely because of successful public awareness programs, wearing cat furs has fallen out of fashion. The United States and Canada still permit the harvest of bobcats and Canada lynx, and Eurasian lynx are harvested in Russia, in each case under regulations that aim to keep the harvest sustainable. The retail cost of a coat of Canada lynx fur ranges from $6,500 to about $17,500 (in 2003 dollars) in the United States, and Eurasian lynx furs from Russia fetch even higher prices. In 2003, the BBC reported that tens of thousands or more domestic cat skins are traded in Europe annually to make coats and other items.

Learning from Cats

In 1981, David H. Hubel and Torsten N. Wiesel won the Nobel Prize in Physiology or Medicine for work that elucidated the structure of the primary visual cortex of the brain. They learned that neurons in this area of the brain fire selectively to stimuli (edges or bars) of differing orientations, and that some of these neurons respond little or not at all to moving edges while others respond strongly to them. They also discovered that there is a critical period in early life for the development of visual pathways from retina to cortex; if stimuli are blocked during this period, permanent visual impairment will result. Hubel and Wiesel did much of their ground-breaking work using domestic cats as subjects. Cats are good models for understanding the human visual system because we share binocular vision. Hubel and Wiesel shared the Nobel Prize with Roger W. Sperry, who also used cats in his studies of the relationship between the two hemispheres of the brain. And six other Nobel Prizes have been awarded to scientists who studied cats to gain a better understanding of the nervous systems.

Although using live animals such as cats in research is controversial, studies like these have considerably advanced our understanding of diverse aspects of human biology. Knowing that calico cats, with their black and orange splotches of fur, are always female and that coat color in cats resides on the X chromosome inspired British geneticist Mary Lyon's discovery that one of the X chromosomes in each cell of a female mammal is turned off, but not every cell turns off the same one. In the calico cat, this leads to a mosaic of cells, some with genes that code for black fur and others that code for orange. This same phenomenon explains a rare skin disorder suffered only by human females.

Domestic cats suffer and die from a virus, feline immunodeficiency virus (FIV), very similar to the HIV virus that causes AIDS in humans, providing the only natural model in which to study this disease. However, Stephen O'Brien and his colleagues found that more than 20 wild cat species carry unique strains of FIV that do not result in fatalities. Reasoning that this might mean that natural selection has occurred for genetic resistance to FIV in the wild cats, O'Brien looked for and found a rare allele, or altered form, of a human gene called CCR5, which makes people resistant to HIV when they are homozygous for this gene (that is, when both chromosomes of a pair carry the rare allele). This, in turn, has suggested a treatment for AIDS. The rare allele codes for a protein that does not permit the virus to raid cells, whereas the protein of the typical alleles lets the virus right in. With this discovery, scientists are looking for ways to give this protection therapeutically to those at risk of AIDS.

Cats on Stage

People use cats for entertainment. Sport hunting of big cats has a long history. The rarity of lions and tigers has made this largely a thing of the past, although some African countries permit sport hunting of lions, leopards, and cheetahs. Hunters are willing to pay handsomely to bag a lion or leopard on an African hunting safari. For example, in 2003 the Tanzanian government charged a $3,000 trophy fee for a lion and $4,000 for a leopard—and that's just the beginning. In that same year, one hunting safari operator offered a 21-day guided hunting safari for $46,450, transportation and other extras not included. In 11 states in the western United States and in 2 Canadian provinces, sport hunting of pumas is legal, and an estimated 2,500 individuals are killed during the puma "seasons" in these areas.

With more people seeking to watch living animals rather than to kill them, the opportunity to see lions or a leopard attracts ecotourists to East Africa, just as tigers lure tourists to India's and Nepal's national parks. In the 1980s, each male lion living in Kenya's Amboseli National Park was estimated to generate about $130,000 annually in tourism revenues. Lions and tigers are always among the most popular attractions in zoos, and circus-goers thrill to the lion- and tiger-tamer acts. For years, until 2003, Siegfried and Roy's magic act featuring white tigers and lions filled more than 1,000 seats, at about $110 each, seven times a week in Las Vegas.

For those who find reading entertaining and rewarding, there is no end of books on cats, mostly about the domestic variety but with a healthy number about their wild relatives as well. A search on Amazon.com for "animal and cat or tiger or lion or leopard or cheetah or puma or jaguar and not fiction" yielded 1,991 book titles.

Cats as Companions

Millions of people find owning and caring for a domestic cat—or being owned *by* one—both entertaining and pleasurable. In the United States about 70 million cats are owned by just about a third of all households. But just as kings and, later, governments amassed menageries of large and exotic cats as evidence of status and empire, some individuals in the United States (and elsewhere) keep wild cats in their backyards, on their rural properties, and even in city apartments. By some estimates, as many as 10,000 or more large cats—mostly lions, tigers, leopards, and pumas—are in private hands in the United States. Young men reportedly purchase puma kittens at state fairs as gifts for their girlfriends. And from time to time, for example, an urban drug dealer is discovered to be the proud owner of a tiger, held to intimidate rivals in that risky business. Very often, these inappropriate pets, perhaps frequently treated as living objects, are abused, depressed, and poorly nourished (zoos feed tigers 2 to 4 kilograms of meat per day; think about the grocery-store cost of purchasing enough meat to properly feed a couple of tigers!).

Many are simply abandoned. Sanctuaries try to care for these animals as best they can, but a better solution would be to ban individual ownership of exotic cats.

WHAT GOOD ARE WILD CATS TO PEOPLE?

Rachel Carson's *Silent Spring,* about the pesticide DDT's deadly impact on songbirds, elicited Aldous Huxley widely circulated lament, "We are losing half the subject-matter of English poetry." Should wild cats go extinct, and many species are perilously close to that, we would lose something very like that. Indeed, it is hard to imagine human culture without cats, so greatly have they influenced our art and literature, our myths and legends, our symbols and psyches, and our history.

Still, it is legitimate to ask what good cats are, especially when their impact on people is sometimes demonstrably bad, as, for instance, when a puma kills a person running through a park in Boulder, Colorado, or a tiger eats a poor villager's cow in India. Conservationists identify four types of arguments that give value to the survival of wild animals such as cats. Two of these arguments justify the value of animals in terms of their contributions to economic prosperity and human survival. These are practical reasons for conservation and for tolerating the occasional tragedy. The first is known as the utilitarian justification: cats are good because they provide us with income. Second is the ecological justification: as an overly simple example, cats are good because without them we would be overrun with rats.

A third argument involves the value of animals' aesthetic appeal: cats are valuable because they are beautiful, inspiring poetry and exciting our imaginations. This is comparable to why we preserve great works of art, and for some people, no more reason than this is required to value cats. Finally, some people believe that cats have a right to exist, just as people have certain rights, and we, as moral beings, are obliged to ensure their survival.

None of these arguments is better than another. People in different times, places, and situations have different needs and values; some people find all of these arguments compelling. Conservationists invoke all of them in their efforts to find ways to protect cats and other wildlife while also ensuring that the lives of people who live near or among them are enhanced.

HOW CAN I BECOME A WILD CAT BIOLOGIST?

Ernst Mayr, a renowned Harvard biologist, said, "Being a biologist does not mean having a job. It means choosing a way of life" (1997, 44). The first thing to do to become a cat biologist is to develop an insatiable curiosity about how the world works, and, in this case, how cats work. You are a biologist first, and you use your biologist's tool chest to enhance our understanding of cat biology.

Then you need a solid undergraduate education in biology, which includes courses in physics, chemistry, genetics, physiology, comparative anatomy, ecology, behavior, calculus, environmental science, English, and history. It is especially important to learn to express yourself in writing and in oral presentations and to understand math thoroughly. Following that, your next step is to choose a graduate school. Your undergraduate professors can help; so, too, can a Web search. Visit with biologists who are using cats as models to answer fundamental questions in biology, including how to keep endangered cat species from going extinct.

You will need to know what specific aspects of cat research you want to work on. Most graduate school professors will not accept you into their research program if you say you simply want to work with cats. You will have to be much more specific than that. Look at your domestic cat, or any cat, to see what more you want to know about it than you already do. Try to look at your cat in ways you never have before, so that you will begin to ask questions about it and thirst for the answers. This is a first step in formulating a significant question that you can answer with your newly acquired biological knowledge and your understanding of the scientific method. Read widely and voraciously to learn what the cutting-edge questions are in the area of biology that most interests you.

If you are still too young to be going off to college or graduate school, you can make a good start by reading this book and simply observing your cat or any cat. Many good books are available on watching cats and explaining their behavior, although, in actual practice, this may be easier read than done. The difficulties involved and the perseverance you will need are well illustrated by the experiences of Paul Leyhausen, author of the classic book *Cat Behavior,* and his assistant R. Wolff in the 1950s.

Leyhausen and Wolff realized that although they had developed a very detailed understanding of cat behavior under laboratory conditions, they knew little about how cats lived in more natural environments. They decided to find out what their domestic cat did every day and every night, how and where it interacted with other neighborhood cats, what prey it captured, and so forth. They planned to work as a team, taking turns following and recording their cat's every behavior and movement as it made its rounds over several weeks on the outskirts of a small German town. They were soon exhausted by their efforts. Leyhausen and Wolff were able to obtain only a very incomplete record of their cat's movement and behavior. Their cat had the upper hand most of the time; they simply could not keep up with it or know where it was on a continuous basis.

Their cat-watching effort, however, did reveal an important cat secret. The domestic cats in their neighborhood seemed never to encounter one another on their forays, even though several cats were using a very small area. On the basis of this observation, and as good biologists, they asked how and why this happened. They developed a working hypothesis: Cats traveling about in the area around the laboratory used space on a time-share plan. Leyhausen and Wolff reasoned that several solitary-living neighborhood cats were able to remain spaced apart in time and never encounter one another by watching from resting places and knowing how long it had been since another cat had passed through, by testing the freshness of another's scent marks, and especially by smelling the preceding cat's urine sprayed on objects such as tufts of grass or the sides of buildings at regularly visited crossing points (see *What Is Scent Marking?*).

Leyhausen and Wolff saw that their neighborhood cats seemed to know who the other cats were and when they were last there, even if they had not seen them. They verified this through repeated observations and found that if a cat came to a marking station another cat had recently passed, it would not continue down the trail after the first cat, but would pause and wait or take a different path. This discovery led to a conceptual change in how biologists explain the ways that solitary-living cats communicate and influence one another's movements while traveling and hunting in the same neighborhood. Leyhausen and his cat-following partner continued their studies of free-living cats and published a major paper on the social organization of solitary-living cats (Leyhausen 1965). This alerted a whole generation of cat biologists to the paradox that solitariness does not preclude social behavior or necessarily diminish its complexity.

Soon cat biologists were watching free-ranging domestic cats living on docks, around human garbage dumps, on farms, and in more remote settings devoid of people, including islands. Other biologists began to look at the other cat species and document how their behavior differs in response to the distribution and abundance of their food resources and what their potential competitors are. They asked, How do these behavior patterns come about? How are they maintained? And why? In the longest-running field study of any wild cat—African lions living in the Serengeti ecosystem—biologists began with the first question: Why do lions live in groups? And from this, new questions have continued to emerge as biologists continually seek to refine their answers. Why do male lions kill some cubs and not others? Who cooperates with whom and why? How did cooperative hunting evolve? Why do individual lions have particular ways of participating in group-territorial conflict? In many instances, after biologists have analyzed repeated, careful, focused observations, the original—and obvious—explanation has turned out to be wrong. This is science at work: never static, always questioning, trying yet one more time for clarification.

Biologists also use cats as model systems or focal species to address a whole host of questions to better understand aspects of nutrition, physiology, reproductive

Capturing and radio-collaring secretive, wide-ranging cats, such as the lynx that scientists have caught in this live trap, enable biologists to learn about their lives in the wild and what they need to survive.

biology, behavioral development, cognition, communication, evolution of social groupings, predator-prey systems, conservation of endangered species, genetics, consequences of inbreeding, and even AIDS. The aim of biologists, and all scientists, is to formulate the principles by which our understanding of the world can be advanced. Knowing ourselves and our biological origins helps us to understand our place in the living world and our responsibility to the rest of nature. Cats have been and will continue to be our partners in this quest.

HOW DO SCIENTISTS STUDY CATS LIVING IN THE WILD?

In the 1800s, specimens of different cat species began arriving in Western natural history museums as collectors traveled the globe in search of material to send back to their own countries. Scientists were in the beginning stages of building a catalog of biological diversity. Just deciding what a "species" is, on the basis of appearances and morphological differences, was exciting and full of controversy—and still is today! Although the killing and collecting of wild cats (or any animal) may be disturbing to us today, at the time, obtaining such material for scientific study was an acceptable, even a noble, pursuit. Much of what we know today about the anatomy of different cat species, their geographical distribution and subspecies, and the relationships among them is based on this earlier collecting as well as on specimens that died in zoological parks and were placed in those museum collections. Today we mostly collect specimens in different ways when we need to, usually in the form of blood or tissue samples and photographs.

Everyone is familiar with domestic cats, and this familiarity creates common knowledge. A second level of knowledge is expert knowledge, the kind a hunter obtains from stalking wild cats or a circus trainer gains over long experience in training lions and tigers to perform. Where wild cats compete for food and kill livestock or people, experts that know the ways of cats are often enlisted to hunt and kill or capture them. Some of our greatest adventure stories were written by these experts in the past century (see *Who Are Some Famous Man-eaters?*). In many cases these hunters were very good observers and accurately recorded their observations. But their writing was also anecdotal and sometimes highly exaggerated. As George Schaller wrote as late as 1967, "The natural history of the tiger has been studied predominately along the sights of a rifle" (1967, 221).

The history of the scientific study of cats living in the wild largely developed in the last half of the twentieth century. With the publication in 1967 of *The Deer and the Tiger,* George Schaller revolutionized the study of wild tigers through the application of the scientific method. He went on to produce masterful field studies of lions, giant pandas, and many other endangered large mammals. Perhaps more important, he showed a whole new generation of cat biologists how to do this work. When Schaller studied tigers in India, his tools were a pair of sturdy walking boots, binoculars, notebooks and pen, a Land Rover, and a certain amount of courage. He carefully and systematically observed tigers through all the seasons of the year. He could identify different individuals, and he charted the course of their different activities. He formulated working hypotheses and supported them with systematic observations. When he was finished, he provided insight into the living, breathing, roaring, hunting, mating, cub-rearing tigers that strongly contrasted with the savage portrait often painted by hunters. His work has been an inspiration to many current and would-be cat biologists. Many other books and papers have now been produced that have enlightened us on the life-history strategies of many of the wild cats. Because most wild cats usually hunt at night, are highly secretive, and the big species can be dangerous, this knowledge has not been gained easily.

To study wild cats in this complicated and complex natural world, scientists ask three big questions: What? How? and Why? *What* questions concern observed traits and behaviors and species composition in an area; *how* questions concern the mechanics that produce those traits, such as physiological processes (e.g., increased hormone levels) and ecological processes (e.g., competition); *why* questions are concerned with ultimate causation: Why did this trait evolve? What selection pressures came to bear to maintain that trait?

Most biologists who study wild cats still living in the wild call themselves ecologists or behavioral ecologists. They want to understand the interactions between cats and their environments. These ecologists look at the ecology of individuals, the ecology of an entire species, and the ecology of communities (that is, the composition and structure of different associated species). Ecologists investigate the precise environmental requirements of particular individuals, such as their climatic tolerance, the resources they need to survive, and their competitors, predators, and diseases. They study the adaptations a species has evolved to live in a particular environment. We assume that every structure of a cat—its morphology, physiology, and behavior—has evolved to be optimal in relation to its environment, including the natural variation in that environment. What concerns conservation biologists today is how much flexibility individual cat species have in responding to environments diminished or changed by human activities.

In investigating the ecology of the different wild cat species at the species level, cat biologists are concerned with the number of individuals that live in a particular area (density), the rate of increase or decrease of that population under varying conditions, and all those factors that control its size, including birth rate, life expectancy, and rates of mortality at different ages. A recent advancement has come through the genetic dimension that ecologists have added; they have begun to examine genetic variation among the individuals in a population and are discovering how this variation is affected by changing environmental conditions, including those resulting from human activities. Ecologists working at the level of the species try to determine what establishes the borders of a species' range and its microdistribution within its geographic range: are the borders and distribution determined by a geological barrier or a limiting factor such as temperature, rainfall, presence of particular vegetation, availability of specific prey, or presence of competitors or other predators?

Biologists are concerned about the dynamics of the metapopulations of a species. In a metapopulation (a set of local populations connected by dispersing individuals), individuals' movements between patches of habitat may be highly restricted by hostile conditions that prevail in these connecting areas. Most wild cat distributions today are fragmented by the activities of people. There is a high probability of extinction in a species that lives in broken populations in small habitat patches that lack connections to one another (see *How Can We Save Wild Cats?* and *What*

Is Happening to All the Wild Cat Habitat?). This is the realm and the study of conservation biology.

Biologists also study wild cats at the community level. They may ask what factors control the number of different cat species that live in a particular place or region? How does this differ by continent or by latitude? The quantities and rates at which energy passes through given ecosystems, sometimes referred to as the food chain, are of particular importance to cats. As predators, cats depend on energy available in the form of prey animals in order to survive, and the amount of this energy available determines the density at which they can live (see *How Do Cat Numbers Affect Prey Numbers and Vice Versa?*).

HOW DO WE TELL INDIVIDUAL WILD CATS APART TO STUDY THEM?

A first challenge in studying the behavioral ecology of wild cats is finding a way to identify individuals in the study population so that we can follow any given one as it goes about its life. In some species such as tigers, leopards, and cheetahs, individuals can be told apart by their markings. Tigers have variable marking patterns on their forehead, either side of the face, legs, and shoulders. K. Ullas Karanth and others have been using these uniquely marked animals to estimate tiger densities in various habitats by photographing them with camera traps. Cameras are set up along pathways, and the tigers take their own photographs by breaking an infrared beam that triggers the cameras (see Figure 6). Two cameras are required because the markings on a tiger's left side vary considerably from those on the right, so simultaneous photos of both sides are needed to be sure the same individual is not being counted twice.

The only field study of serval ecology and behavior was possible because Aadje Geertsema (1985) was able to recognize the variation of stripe and spot patterns in 34 individual animals. Lion biologists have devised a way to recognize individual lions on the basis of the variation in the mystacial vibrissa spots (the pattern of spots associated with their muzzle whiskers). To identify individuals and follow the breeding history of lions living in the isolated Ngorongoro Crater in Tanzania, Craig Packer (1992) obtained photographs of them taken over the years by tourists who had visited since a 1962 biting fly epidemic reduced lion numbers to a handful. The tips of cheetahs' tails differ on each side and also from one individual to another. Tim Caro and Sarah Durant (1991) have photographed hundreds of cheetah tails in the Serengeti and read them as they would a bar code at a grocery store checkout. Using a computer-aided matching system for coat patterns on cheetah

Variation in the pattern of whisker spots uniquely identifies individual lions just as fingerprints do people. Stripe patterns do the same for tigers, as do the tail bars of cheetahs and the spots and stripes of servals. The ability to identify individuals greatly enhances scientists' accuracy in censusing the number of cats in an area and helps them to determine such life-history parameters as longevity and reproductive success.

flanks, Marcella Kelly (2001) has identified individuals with 97 percent accuracy from 10,000 photographs of Serengeti cheetahs taken over 25 years.

Over the years a great deal of controversy has arisen about an expert's ability to tell individual big cats apart by their tracks—called pug marks in India. Under the right conditions, in some areas biologists have been tested and have shown that they can, for the most part, accurately differentiate among the tracks of individual tigers, for example. But other biologists studying pumas prefer not to rely on the discriminating eye and the exceptional inherent ability of some biologists to match patterns and instead use a set of track measurements to differentiate individuals. Tracks are photographed, scanned, digitized, and the measurements compared using GIS (geographic information system) technology. Indian biologists are exploring this method to identify Asian lions and tigers by their tracks.

Figure 6. Scientists often set out cameras that take a photograph whenever a cat, or other animal, breaks an infrared beam projecting across a path. This enables the scientists to determine the distribution and abundance of a species in a particular area. Individual cats that appear in the photographs are identified by their unique stripe or spot patterns. (Drawing by P. A. Miththapala)

Some dogs can be readily trained to differentiate between the tracks of their target species and those they should ignore of other animals that cross their scent trail. Other biologists, such as those studying tigers, are now taking this one step further and training dogs to differentiate individual tigers by their feces, or scats, as biologists call them. Biologists search an area and gather samples of tiger feces from different sites. These samples are taken back to camp, where the trained dogs indicate which scats are from the same or different tigers. Dogs could be used to search out scats of some cat species, especially those cats that are difficult to locate, but with tigers this is not a good idea, because tigers catch and eat dogs when they can.

Molecular scatology is a recent advancement. Molecular techniques have greatly increased the accuracy of identifying prey in scats. In addition, the sex, age, reproductive status, and physiological stress level of the cat depositing the scat can be ascertained by looking at fecal steroid hormones. Feces also contain cells shed from the lining of the cat's intestine, from which biologists isolate, purify, and amplify specific DNA sequences. DNA can be used to recognize different species and individuals and determine their sex. It can also be used to determine whether all the cats living in an area are related to one another and how. This allows biologists to determine levels of genetic variation in fragmented populations, rates of dispersal among different populations, and when dispersal has ceased and a population has become isolated. Once we have the ability to identify individuals through their scats, we can determine how many young a female has and who the father is. It allows us to census the size of the population and to determine territory size.

This same information can be obtained from DNA isolated from a cat's hair follicles. And how do biologists capture just the hair from a cat for DNA analysis? One ingenious method was devised by David Smith and Charles McDougal and their assistants, who also first enabled a tiger to photograph itself with a camera trap. A passing cat gives hair samples by tripping a trigger that releases a light bamboo cane with sticky flypaper affixed to the end. The cane lightly swats the cat, and the flypaper snares a few hairs. Other biologists use catnip or other lures to induce wild cats to rub against a small carpet shag or another hair-sticking device to leave a few hair samples.

The same molecular genetic techniques that allow us to identify individual cats also help to identify prey in a cat's diet and to assess levels of a cat's parasite infestations. These techniques are particularly useful because in many situations prey choice cannot be established or quantified using direct observation, because the bone fragment, teeth, and hair of the animals the cat ate are not quantifiable or, in many cases, even identifiable.

HOW ARE BIG CATS CAUGHT?

John has had the privilege of studying the behavior and ecology of lions, tigers, leopards, pumas, fishing cats, and bobcats in their native habitats. Because most cats live at low densities, keep to thick cover, and move mostly at night, direct observation over continuous periods is usually not possible. After a break in observation, relocating your focal study animals can take a lot of time if they are cats that spend much of their time in large expanses of open grasslands and savannas, such as lions and cheetahs.

In these habitats, biologists place a radio-transmitter collar on a focal animal, such as an older lioness in a pride, so that the pride can be easily located. In this way more of the biologist's time can be spent watching lions rather than trying to find them. When the lions are used to people in Land Rovers, biologists can just drive up to the pride and dart a lion with an anesthetic-filled dart. Where lions have not been so habituated, stereo boom-box speakers are used to broadcast tape recordings of roaring and of lions and hyenas in feeding skirmishes and making calls at a kill. The resident pride members respond by coming to the location of the boom box, where they can then be darted or first snared and then darted.

In John's study of puma social organization in the Salmon River Mountains of Idaho, our task was to capture and place radio collars on all the adult pumas living in this wilderness area. Deep snow in the higher elevations forced the puma's principal prey, mule deer and elk, down to lower elevations, where there was less snow in the late fall. The pumas followed the deer and elk down, and this is where they were captured. But captured how? In their evolutionary history, pumas were pursued and killed by another predator, probably wolves. Hunters, and now biologists, have found that it doesn't take much pursuing by a barking dog to send a puma up a tree. To capture a puma, we walked up and down creeks and canyons until we found a fresh puma kill or saw fresh puma tracks in the snow. Knowing a puma was nearby, we released a specially trained hound or two. The hounds bounded after the cat, following its scent trail. Because of their training, they were not deterred by other scents they encountered in pursuit of the cat. Usually, the baying hound rapidly caught up with the puma, and the puma climbed into a tree—usually a tree that was very difficult for us to climb. The hound circled the base of the tree, barking and baying, preventing the puma from escaping before we could reach the tree with our backpacks filled with equipment. We then darted the puma with an anesthetic drug. The drug acted slowly enough that the puma could cling to the tree for a few minutes before going completely under, giving us enough time to climb the tree with pole-climbing spurs (the kind that telephone linemen use), attach a rope to the puma's legs, and lower it to the ground, where we attached a radio transmitter before the cat recovered.

In a variation on this method, biologists trying to catch Florida panthers tree a panther with hounds, and before they dart it, they spread a specially designed air mattress on the ground below to catch the panther and break its fall from the tree when the anesthesia takes effect. Hounds have been used very effectively to capture bobcats also but do not usually work as well in treeing other big cats such as leopards, jaguars, and tigers, which seem to have little inherent fear of dogs.

In Nepal, John employed trained elephants, an ancient technology, to catch tigers and place radio collars on them. Riding trained elephants, we searched through the tall grass to find a fresh tiger kill. In a few instances, the tiger was darted from the back of the trained elephant. Alternatively, we darted the tiger from the safety of nearby trees after climbing into them and awaiting the tiger's return to the kill.

Other biologists used another old hunting technique. After a tiger was located with a kill, a surrounding area of several acres was ringed with a white cloth fence, the great length of cloth being unloaded from the backs of trained elephants. For some reason, even though this was a narrow band of cloth, only a few feet high, tigers did not want to jump it or rush through it to escape, which they could easily have done.

The biologists left a narrow opening in the ring near a stand of trees, then entered it on trained elephants, and slowly pushed the tiger toward the opening, where other biologists with dart guns were stationed in the trees. When the elephants moved forward slowly and deliberately, the tiger had time to test the perimeter of the cloth ring, detect the opening, and try to make its escape. If the elephants pushed forward too fast, the tiger would turn and attack them. As the tiger made its "escape" through the opening in the cloth fence, it could be darted by one of the biologists waiting above in the trees.

Trained elephants are not available in much of the tiger's range. Where this is the case, biologists have caught tigers and leopards with large box traps constructed of logs or heavy-gauge woven wire. A problem with box traps is that, once captured in this manner, an animal usually becomes immediately "trap shy" and is not easily enticed into this kind of trap again. A few inexperienced biologists, in a hurry to get their projects underway, began their box-trapping with considerable success, but before their radio-tracking equipment had cleared customs. When the radio-tracking equipment finally became available, they found that their potential study animals were studiously avoiding the box traps they so readily had entered earlier.

Wire-mesh box traps set for one species have also succeeded in capturing cats other than the intended ones. Wire box traps set for leopard cats in Thailand forests, for example, netted a leopard and two clouded leopards. It is a wonder that a leopard or a clouded leopard could even fit into the wire-mesh box trap baited for the smaller leopard cat. Nearly all we know about the movements and activities of wild clouded leopards has come as a result of these fortuitous captures. John has also captured bobcats in wire box traps set for raccoons.

Poachers frequently use snares to capture cats to kill them. This very old technology has been made more deadly with the advent of light, flexible wire cable. Poachers set these snares so that the cat is captured by the neck and strangled. Bear biologists invented a modified snare that throws the snare cable loop high around the bear's leg when the bear steps in it. Thousands of bears have now been captured for study in this way. Recently, these modified foot-hold snares have been used successfully to capture tigers, lions, pumas, and lynx. But you cannot use a snare in a situation where the snared animal can be killed by another animal; there is great concern, for instance, that elephants would kill tigers or leopards caught in snares. There is also a danger of a snared animal suffering from high daytime temperatures or below-freezing temperatures. To avoid injury to a study animal, snares are checked frequently and are usually equipped with a radio transmitter that begins to

Much remains to be learned about all wild cats, from majestic tigers to tiny kodkod, if we are to ensure the conservation of all they need to survive in the wild. In scientists' search for the knowledge that will save these extraordinary animals for our grandchildren to marvel at, they are using a combination of traditional methods, such as observation from elephant-back or from a blind in the forest, and new, high-tech methods, such as camera-trapping and analyzing DNA extracted from feces.

transmit a signal to alert the biologist when they have been sprung. Biologists place snares along trails or attract passing cats to investigate with a scent lure.

Other ingenious means of attracting cats to snares have been used. Lions respond to the sounds of squabbling over a kill, as noted above, and the taped calls of a stressed buffalo calf. Tigers respond to the distress calls of a wild piglet played over a boom box. Alternatively, biologists find a natural kill and surround it with several snares. As you can see, biologists use mix of ancient and modern technologies to capture cats in their quest to learn more about how cats live in the wild.

WHAT IS LEFT TO LEARN ABOUT CATS?

Remarking on the history of ideas, Oxford scholar Theodore Zeldin concluded, "The really big scientific revolutions have been the invention not of some new machine, but of new ways of talking about things" (2000, 83). Where do new ideas and new ways of seeing and talking about things come from? As biologists, we are always searching for new ways to bring events, processes, and observations into a focus we can understand. Our minds don't create the patterns we see in nature, but we must try to see them, or any familiar phenomena, in novel ways and strive for clarity. Leyhausen did this when he tried to understand his domestic cat's free-ranging behavior in minute detail (see *How Can I Become a Wild Cat Biologist?*). The distinguished evolutionary biologist Niles Eldredge calls this "making the world more visible" (2000, 16). Eldredge points out that the scientific process is about matching up our mental pictures with the biological and physical reality in the world. It is a search for the absolutely best and most accurate description of the world and how it works. Our great life adventure as biologists is to learn to see biological patterns in new and ever more accurate ways.

Take the seemingly simple question, How many cat species are there? For the moment, 40 cat species are recognized and listed in the most recent edition of *Mammal Species of the World,* a compilation of species agreed upon by the world's experts. Initially, the determination of species was based on morphological analysis, but this analysis of the cat specimens available in museums has advanced our understanding about as far as it can. Exciting new biochemical methods can now further clarify and advance our understanding of the evolutionary relationships among cats and the patterns in their diversity. Molecular tools have recently yielded results that strongly suggest deep divisions between some species, divisions that were previously undetected through morphological analysis. As these recent findings are included in our thinking, no doubt additional cat species will be recognized. We continue to seek to understand how these patterns in the diversity among the felids match our understanding of the template of the physical and biological world in which they have evolved. In this case, new tools are important, but just as important are new ways to think about these new findings.

Human activities are causing massive environmental changes in the world, prompting us to ask, How can we secure the future of all the wild cats in the wild? To accomplish this we need a thorough understanding of each species' behavioral and ecological needs. For example, we need to understand the flexibility in their mating, rearing, dispersal, foraging, and refuging behavioral systems, each wild cat's risk of extinction under varying mortality schedules, what limits their geographical distribution, and how the range of each species is contracting. Certain entire species—and certain populations within *all* species—are at risk of extinction, yet we know virtually nothing about them. Browsing through the recent encyclopedic compilation *Wild Cats of the World* (Sunquist and Sunquist 2002) reveals that the number of pages written about what we know of each species directly relates to the species' body size, with the exception, perhaps, of the lynxes. For most of the smaller species, virtually nothing is known beyond their description and an account of how they were first discovered by Western biologists. For others, even widespread species, nothing much beyond a few studies of food habits have been made, and these may not be representative. As molecular techniques have been perfected, biologists have gone back to examine museum specimens and collect samples from them for molecular analysis, and they have found that some study skins of the very rare species were misidentified. The lesson: We have even less information than we thought, and some of what we have is misleading, even wrong.

Molecular biology involving cats is also providing powerful new tools to expand our understanding of other species. The Feline Genome Project, being undertaken at the National Cancer Institute in the Laboratory of Genomic Diversity, involves a compilation of a gene map and eventually the whole genome sequence of the domestic cat. Even before it is completed, this will be of great value for understanding the human genome. It will provide insights into human hereditary diseases and infectious diseases, improve veterinary medicine, and even be used in forensics. It will also enable us to better understand the evolution of genome organization among mammals, adaptive evolutionary divergence, and conservation genetics.

WHERE CAN I GET MORE INFORMATION ABOUT CATS?

Powerful Web search engines allow nearly instant access to information that was formerly available only to specialists with access to a large library. But information found on the Web isn't necessarily correct or up to date. Anyone can post material on the Web, and usually it has not been checked for accuracy, or peer reviewed, by professionals. However, many useful Web sites are monitored for accuracy if not formally peer reviewed (see appendix 3).

Biologists and students are contributing new information gathered from ongoing studies of cats, and the results are available online as the studies progress. In a pioneering partnership, biologists with the Washington Department of Fish and Wildlife and the Hornocker Wildlife Institute of the Wildlife Conservation Society joined local students from kindergarten through the 12th grade in a project called CAT (Courage and Teaching). As a team, they are tracking puma movement 24 hours a day for 8 years through the use of Global Positioning System (GPS) collars. The biologists attach the collars to the pumas, which are captured with snares (see *How Are Big Cats Caught?*). Information on puma movements is gathered by the biologists, and the students then download the information and, under the watchful eyes of the biologists, plot it onto forest classification maps. This team is learning how pumas are using their environment and how wildlife managers and community planners can design communities compatible to both wildlife and humans in areas where hundreds of new houses are slated for construction in puma habitat. Not only do the students help in the analysis of the data, but also they become teachers themselves and go out and tell the community about their findings, helping others to learn how to live with great cats in their midst.

APPENDIX 1.

SCIENTIFIC AND COMMON NAMES OF THE FAMILY FELIDAE

Acinonyx jubatus	cheetah
Caracal caracal	caracal
Catopuma badia	bay cat
Catopuma temminckii	Asian golden cat
Felis bieti	Chinese mountain cat
Felis catus	domestic cat
Felis chaus	jungle cat
Felis manul	Pallas' cat
Felis margarita	sand cat
Felis nigripes	black-footed cat
Felis silvestris	wildcat
Leopardus braccatus	Pantanal cat
Leopardus colocolo	Chilean pampas cat
Leopardus geoffroyi	Geoffroy's cat
Leopardus guigna	kodkod
Leopardus jacobitus	Andean mountain cat
Leopardus pajeros	Argentinean pampas cat
Leopardus pardalis	ocelot
Leopardus tigrinus	oncilla
Leopardus wiedii	margay
Leptailurus serval	serval
Lynx canadensis	Canada lynx
Lynx lynx	Eurasian lynx
Lynx pardinus	Iberian lynx
Lynx rufus	bobcat
Neofelis nebulosa	clouded leopard
Panthera leo	lion
Panthera onca	jaguar
Panthera pardus	leopard
Panthera tigris	tiger
Pardofelis marmorata	marbled cat

Prionailurus bengalensis	leopard cat
Prionailurus iriomotensis	Iriomote cat
Prionailurus planiceps	flat-headed cat
Prionailurus rubiginosus	rusty-spotted cat
Prionailurus viverrinus	fishing cat
Profelis aurata	African golden cat
Puma concolor	puma
Puma yagouaroundi	jaguarundi
Uncia uncia	snow leopard

APPENDIX 2.

CONSERVATION STATUS OF CATS OF THE WORLD IN 2003, ACCORDING TO TOP THREE AUTHORITIES

		CITES Appendixes†	IUCN Red List	US Endangered Species List
Cheetah	*Acinonyx jubatus*	I	Vulnerable	Endangered
Puma	*Puma concolor*	I eastern US; II otherwise	Near threatened	Endangered in eastern US
Jaguarundi	*Puma yagouaroundi*	I North and Central America; II otherwise	Least concern	Endangered in North and Central America
Iberian lynx	*Lynx pardinus*	I	Critically endangered	Endangered
Eurasian lynx	*Lynx lynx*	II	Near threatened	
Canada lynx	*Lynx canadensis*	II	Least concern	Threatened in US
Bobcat	*Lynx rufus*	II	Least concern	Endangered in Mexico
Clouded leopard	*Neofelis nebulosa*	I	Vulnerable	Endangered
Snow leopard	*Uncia uncia*	I	Endangered	Endangered
Lion	*Panthera leo*	I in India; II in Africa	Vulnerable	Endangered in India
Tiger	*Panthera tigris*	I	Endangered	Endangered
Leopard	*Panthera pardus*	I	Least concern	Endangered in Asia and northern Africa
Jaguar	*Panthera onca*	I	Near threatened	Endangered
Leopard cat	*Prionailurus bengalensis*	I India, Bangladesh, Thailand; II otherwise	Least concern	Endangered in Thailand, India, Bangladesh
Iriomote cat	*Prionailurus iriomotensis*	II	Endangered	Endangered
Fishing cat	*Prionailurus viverrinus*	II	Vulnerable	
Flat-headed cat	*Prionailurus planiceps*	I	Vulnerable	Endangered

		CITES Appendixes[†]	IUCN Red List	US Endangered Species List
Caracal	*Caracal caracal*	I Asia; II otherwise	Least concern	Endangered
African golden cat	*Profelis aurata*	II	Vulnerable	
Bay cat	*Catopuma badia*	II	Endangered	
Asian golden cat	*Catopuma temminckii*	I	Vulnerable	Endangered
Marbled cat	*Pardofelis marmorata*	I	Vulnerable	Endangered
Rusty-spotted cat	*Prionailurus rubiginosus*	I India; II Sri Lanka	Vulnerable	
Serval	*Leptailurus serval*	II	Vulnerable	Endangered
Domestic cat	*Felis catus*			
Wildcat	*Felis silvestris*	II	Least concern	
Chinese mountain cat	*Felis bieti*	II	Vulnerable	
Sand cat	*Felis margarita*	II	Near threatened	Endangered
Jungle cat	*Felis chaus*	II	Least concern	
Black-footed cat	*Felis nigripes*	I	Vulnerable	Endangered
Pallas' cat	*Felis manul*	II	Near threatened	
Andean mountain cat	*Leopardus jacobitus*	I	Endangered	Endangered
Ocelot	*Leopardus pardalis*	I	Least concern	Endangered
Margay	*Leopardus wiedii*	I	Least concern	Endangered
Geoffroy's cat	*Leopardus geoffroyi*	II	Near threatened	
Kodkod	*Leopardus guigna*	II	Vulnerable	
Oncilla	*Leopardus tigrinus*	I	Near threatened	Endangered
Pantanal cat	*Leopardus braccatus*	II	Near threatened	
Chilean pampas cat	*Leopardus colocolo*	II	Near threatened	
Argentinean pampas cat	*Leopardus pajeros*	II	Near threatened	

[†]Appendix I is the highest category of threat; trade in these species threatens them with extinction, so is severely restricted. Appendix II is the category for species whose trade must be based on sustainable harvests.

APPENDIX 3.

WEB RESOURCES ON CATS

African lion research—University of Minnesota: www.lionresearch.org

Carnivore Conservation: www.carnivoreconservation.org

Cat Action Treasury (CAT): www.felidae.org

Cat Specialist Group (IUCN—The World Conservation Union): www.catsg.org

Cheetah Conservation Fund: www.cheetah.org

CITES—Convention on International Trade in Endangered Species of Wild Fauna and
Flora: www.cites.org

Clouded leopard project: www.cloudedleopard.org

Felid Taxon Advisory Group—American Zoo and Aquarium Association: www.felidtag.org

International Snow Leopard Trust: www.snowleopard.org

IUCN—The World Conservation Union Red List of Threatened Species: www.redlist.org

Jaguar conservation—Wildlife Conservation Society: www.savethejaguar.org

Mountain Lion Foundation: www.mountainlion.org

Project Tiger, India: www.enfor.nic/pt

Smithsonian's National Zoological Park: http://nationalzoo.si.edu

Snow Leopard Conservancy: www.snowleopardconservancy.org

Tiger Information Center—Save The Tiger Fund: www.5tigers.org

Tigris Foundation: www.tigrisfoundation.nl

TRAFFIC—Wildlife trade monitoring organization sponsored by IUCN and WWF:
www.traffic.org

Wildlife Conservation Society: www.wcs.org

GLOSSARY

adaptations A set of features—anatomical structures as well as behaviors—enabling animals to survive and reproduce successfully in the habitats in which they live.

allometry How the proportions of various parts of an animal change as the animal's size changes.

altricial young Young born helpless, naked, blind, and incapable of survival independent of their mother for a relatively long time after their birth. Altricial and its opposite, precocial, are actually the ends of a continuum that describes the relative maturity of animals at the time of their birth. Compare **precocial**.

arboreal Living in trees; adapted to life in trees. Compare **terrestrial**.

basal metabolic rate The rate at which energy is used by an animal at complete rest. Basal metabolism is the minimum amount of energy needed to maintain vital functions in an awake animal at complete rest in a comfortably warm place.

binocular vision The ability to focus both eyes at once on an object. This ability enables cats to pounce on prey accurately.

biomass The total weight of the organisms or of a particular group or type of organism living in a particular area.

birds of prey Flesh-eating birds, such as hawks, eagles, and owls.

boreal Of or relating to the forests of the northern North Temperate Zone, dominated by coniferous trees such as spruce, fir, and pine.

caching Storing food so that it can be eaten at a later time.

camouflage The way an animals blends into its environment in order to sneak up on prey and hide from predators. The coats of most cats help them to blend into the habitat in which they live and hunt.

canid A member of the family Canidae, which includes dogs, wolves, and foxes.

canine teeth The teeth between the front incisor teeth and the side molars. Long, sharp canine teeth are a feature of all cats, which use them to kill prey.

carnassials Scissorlike premolar and molar teeth that tear chunks of meat from a carcass.

Carnivora An order of mammals, most of which are meat-eaters, including cats, dogs, bears, and others.

carnivoran A member of the order Carnivora; used to distinguish these member species from other meat-eaters, or carnivores. Compare **carnivore**.

carnivore Any animal that eats meat or flesh. Many animals, including humans and eagles, are carnivores but do not belong to the order Carnivora. Compare **herbivore**.

clade A group of animals whose members all derived from a common ancestor; also called a lineage. The family Felidae (cats) is a clade, and each lineage within the cats is also a clade.

closed habitats Areas with a significant amount of plant cover, such as forests. Compare **open habitat**s.

coalition A small group of male cheetahs or lions, often brothers, that live, hunt, and defend a territory together.

competitors Two or more animals that may fight for the same food, territory, habitat, or mating partners.

conservation The effort to maintain the earth's natural resources, including wildlife, for future generations.

convergent evolution Similar features or behaviors evolved independently by distantly related animals. The saber teeth of saber-toothed cats and saber-toothed marsupials are an example of convergent evolution.

crepuscular Active at dawn and dusk. Compare **diurnal** and **nocturnal**.

desert Dry region receiving less than 250 millimeters of rain annually. Desert-living cats include sand cats and black-footed cats.

digitigrade stance Walking on the toes, so that the heels do not touch the ground. The foot bones of cats are modified so that only their toes touch the ground. Human stance, walking with the foot flat on the ground, is called plantigrade.

diurnal Active during the day. Compare **crepuscular** and **nocturnal**.

dispersal The process of an animal leaving the place of its birth to find a home range elsewhere.

endangered species Species that are likely to die out (become extinct) unless people take action to prevent this. In casual usage, many species are called endangered, but the word has a legal and scientific definition under the US Endangered Species Act and a strict scientific definition on the IUCN Red List of Threatened Species.

estrus The period when a female mammal is ready to copulate with a male. Estrus usually occurs around the time a female ovulates, or produces an egg or eggs that can be fertilized by male sperm.

extinct No longer living. When the last living member of a species dies, the species is extinct.

felid A member of the family Felidae, or cats.

flagship species Flagship species are popular, charismatic species that serve as symbols and rallying points to stimulate conservation awareness and action. Tigers and giant pandas are flagship species.

flehmen A behavior in which cats curl their lips into a grimace after sniffing another cat or its scent mark. Flehmen may serve to draw chemicals into the vomeronasal organ. See **vomeronasal organ**.

genome An animal's genetic material.

habitat The place where an animal lives in the wild, such as a forest or grassland. An animal's habitat provides food, water, shelter, potential mates, and the right environment for the animal's survival.

herbivore An animals that eats primarily plant material. Many of the species that cats prey on, such as deer, antelope, and rabbits, are herbivores. Compare **carnivore**.

home range The area that an animal travels over during the course of a year to find food and shelter, to find mates, and to rear young. Compare **territory**.

hypercarnivores Animals that eat only meat. Cats are hypercarnivores.

incisor teeth The front teeth. Cats use their incisor teeth for fine work such as plucking feathers from a bird carcass.

indicator species Species whose presence, absence, or a change in their distribution and abundance reflects changes in environmental quality or other measures. For example, in environments where the seasons are marked, changes in the time of flowering of some plant species or in the birth season of mammals may indicate that the climate has changed.

infanticide Killing young animals. In biology, this usually refers to an animal killing young of its same species (killing the young of another species is called predation).

keystone species A species that through its size, activity, or productivity has an impact on its community or ecosystem greater than expected on the basis of its relative abundance. The real impact of a keystone species is usually detected when the species is no longer present.

kleptoparasitism One animal stealing food from another animal.

lagomorph A member of the order Lagomorpha, which includes rabbits, hares, and pikas, primary prey animals of some cats, especially of lynx species.

landscape species Species whose conservation requires the conservation of entire large regional landscapes to meet their ecological needs. Tigers are a landscape species.

lean-season prey biomass The combined weight of all potential prey available when food for prey is least abundant.

leucism Light or white coloration of skin, hair, fur, and feathers, such as the white fur of white tigers and the light fur of domestic Siamese cats. Compare **melanism**.

lineage A group of animals whose members all derived from a common ancestor; also called a clade.

melanism Dark or black coloration of skin, hair, fur, and feathers. Melanistic cats, such as black leopards, have dark or black fur. Compare **leucism**.

nocturnal Active at night. Compare **crepuscular** and **diurnal**.

olfaction The sense of smell.

open habitats Areas with relatively little plant cover, such as savannas and deserts. Compare **closed habitats**.

pheromones Chemicals produced by animals that send a message to others of the same species, especially chemicals that convey information related to reproduction.

precocial young Animals born at a relatively advanced state of maturity and capable of survival independent of their mother very soon after birth. Precocial and its opposite, altricial, are actually the ends of a continuum that describes the relative maturity of animals at the time of their birth. Compare **altricial young**.

poaching Hunting animals illegally.

predator An animal that hunts, kills, and eats other animals to survive. All cats are predators.

prey Animals that are hunted, killed, and eaten by other animals called predators.

pride A group of female lions and their young. The females in a pride are usually related. Males attach themselves to prides of females.

primate A member of the order Primates, which includes monkeys and apes (of which humans are one species).

rain forest A forest that receives at least 2.5 meters of rainfall annually. Most rain forests are in tropical regions of the world. Some tigers and leopards live in rain forest.

riparian Relating to the banks of a water course such as a river or stream. Fishing cats occupy riparian habitats.

rodent A member of the order Rodentia, which includes rats and mice. Rodents are the primary prey of most small cats.

scavenger An animal that eats meat killed by predators.

savanna An area with relatively sparse mixed vegetation of grasses, trees, and shrubs. Lions are cats of the African savanna.

species A group of animals with very similar features. Individual members of a species are able to breed and produce viable young that are fertile (able to breed when they too become adults). Under usual natural conditions, individuals of different species do not interbreed, but some exceptions occur. The species is the basic unit in scientific classification of animals and plants.

subspecies Populations of a species in which individuals consistently differ in certain features from individuals in other populations of that species but not enough that the individuals can no longer breed and produce viable, fertile young. Subspecies are usually separated from one another by barriers such as seas or high mountains. If subspecies remain isolated from one another for a very long time, they may become different species.

standing crop biomass The total weight of the prey available to a predator at any one time in a defined area.

tapetum lucidum A structure in a cat's eye that reflects light back through the eye so that the cells responding to light get a second chance to respond to and use that light. Many nocturnal mammals and most carnivorans have this structure. This is what causes the "eyeshine" of cats when they look into a light at night.

taxon (sing.), taxa (pl.) A category or group of animals. A taxon may be a species, genus, family, or other taxonomic group.

terrestrial Living on land; adapted to living on land. Compare **arboreal**.

territory A home range that an animal (or group of animals of the same species) lives in and defends from other members of its species, especially those of the same sex.

trophic pyramid A term that summarizes the observation that meat-eating predators, at the top of the pyramid, are less abundant than plant-eating prey, which in turn are less abundant than plants, because the higher up the pyramid, the smaller the amount of energy available. As a rule of thumb but with some variation, about 10 percent of the sun's energy is fixed by photosynthesis into green plants, and about 10 percent of that energy passes to herbivores, and about 10 percent of herbivore energy is available to carnivores.

umbrella species A species whose areas of occupancy or home ranges are large enough and its habitat requirements are wide enough—or exacting enough—that setting aside a sufficiently large area for its protection will consequently protect many other species as well.

ungulates Large, plant-eating mammals with hooves (toes encased in thick, hard skin and having sharp edges). They include elephants, rhinoceroses, horses, deer, antelope, wild cattle, and their relatives. Ungulates are the primary prey of large cats.

vibrissae Long stiff hairs, or whiskers, that project from the face and other parts of a cat's body. They are sensitive to the lightest touch.

vomeronasal organ Two tiny openings on the roof of the mouth through which chemicals send messages to parts of the brain concerned with sexual behavior; also known as the Jacobsen's organ. See **flehmen**.

REFERENCES

GENERAL REFERENCES

Ewer, R. F. 1973. *The Carnivores*. Ithaca: Cornell University Press.

Gittleman, J. L., ed. 1989. *Carnivore Behavior, Ecology, and Evolution*. Ithaca: Cornell University Press.

———, ed. 1996. *Carnivore Behavior, Ecology, and Evolution*, vol. 2. Ithaca: Cornell University Press.

Guggisberg, C. 1975. *Wild Cats of the World*. New York: Taplinger Press.

Kitchener, A. 1991. *The Natural History of the Wild Cats*. Ithaca: Cornell University Press.

Leyhausen, P. 1979. *Cat Behavior*. New York: Garland STPM Press.

Lumpkin, S. 1993a. *Big Cats*. New York: Facts on File.

———. 1993b. *Small Cats*. New York: Facts on File.

Macdonald, D. 1992. *The Velvet Claw: Natural History of the Carnivores*. London: BBC Books.

Morris, D. 1994. *Illustrated Catwatching*. New York: Crescent Books.

———. 1996. *Cat World, a Feline Encyclopedia*. New York: Penguin.

Neff, N. 1986. *The Big Cats*. New York: Abrams.

Nowell, K., and P. Jackson. 1996. *Wild Cats*. Gland, Switzerland: International Union for the Conservation of Nature and Natural Resources.

Seidensticker, J., and S. Lumpkin, eds. 1991. *Great Cats*. Emmaus, PA: Rodale Press.

Sunquist, M., and F. Sunquist. 2002. *Wild Cats of the World*. Chicago: University of Chicago Press.

Turner, D. C., and P. Bateson, eds. 2000. *The Domestic Cat: The Biology of Its Behaviour*. 2d ed. Cambridge: University of Cambridge Press.

Wolfe, A., and B. Sleeper. 1995. *Wild Cats of the World*. New York: Crown Publishers.

REFERENCES BY SUBJECT

Preface

Tinbergen, N. 1970. *The Herring Gull's World*. Rev. ed. New York: Harper and Row.

1. Cat Facts

Classification

Bininda-Emonds, O. R. P., J. L. Gittleman, and A. Purvis. 1999. Building large trees by comparing phylogenetic information: A complete phylogeny of the extant Carnivora (Mammalia). *Biological Review* 74: 143–175.

Savage, R. J. G. 1977. Evolution in carnivorous mammals. *Palaeontology* 20: 237–271.

Wozencraft, C. W. In press. Family Felidae. In D. E. Wilson and D. M. Reeder, eds., *Mammal Species of the World*. 3d ed. Washington, DC: Smithsonian Books.

Physical Features and Senses

Alexander, R. M. 2003. *Principles of Animal Locomotion*. Princeton: Princeton University Press.

Blumberg, M. S. 2002. *Body Heat: Temperature and Life on Earth*. Cambridge, MA: Harvard University Press.

Cates, S. E., and J. L. Gittleman. 1997. Reading between the lines—is allometric scaling useful? *TREE* 12: 338–339.

Davis, D. D. 1962. Allometric relationships in lions vs. domestic cats. *Evolution* 16: 505–514.

Eisenberg, J. F. 1986. Life history strategies of the Felidae: Variation on a common theme. Pp. 293–303 in S. D. Miller and D. D. Everett, eds., *Cats of the World: Biology, Conservation, and Management*. Washington, DC: National Wildlife Federation.

Eizirik, E., and S. J. O'Brien. 2003. Evolution of melanism in the Felidae. *Cat News* 38: 37–39.

Eizirik, E., et al. 2003. Molecular genetics and melanism in the cat family. *Current Biology* 13: 448–453.

Emmons, L. H. 1987. Comparative feeding ecology of felids in neotropical rainforest. *Behavioral Ecology and Sociobiology* 20: 271–283.

Gittleman, J. L. 1985. Carnivore body size: Ecological and taxonomic correlates. *Oecologia* 67: 540–554.

———. 1991. Carnivore olfactory bulb size: Allometry, phylogeny and ecology. *Journal of Zoology, London* 225: 253–272.

Godfrey, D., J. N. Lythgoe, and D. A. Rumball. 1987. Zebra stripes and tiger stripes: The spatial frequency distribution of the pattern compared to that of the background is significant in displays and crypsis. *Biological Journal of the Linnean Society* 32: 427–433.

Gonyea, W. J. 1976. Adaptive differences in the body proportions of large felids. *Acta Anatomy* 96: 81–96.

Gonyea, W., and R. Ashworth. 1975. The form and function of retractile claws in the Felidae and other representative carnivores. *Journal of Morphology* 145: 229–238.

Grand, T. I. 1976. The anatomical basis of locomotion. *American Biology Teacher* 38: 150–156.

Guillery, R. W. 1974. Visual pathways in albinos. *Scientific American* 230(5): 44–54.

Hedrick, A. V., and E. J. Temeles. 1989. The evolution of sexual dimorphism in animals: Hypothesis and tests. *TREE* 4: 136–138.

Hoogesteijn, R., and E. Mondolfi. 1996. Body mass and skull measurement in four jaguar populations and observations on their prey base. *Bulletin of the Florida Museum of Natural History, Biological Series* 39: 195–219.

Huang, G. T., J. J. Rosowski, and W. T. Peake. 2000. Relating middle-ear performance to body size in the cat family: Measurements and models. *Journal of Comparative Physiology A* 186: 447–465.

Huang, G. T., et al. 2002. Mammalian ear specializations in arid habitats: Structural and functional evidence from sand cat (*Felis margarita*). *Journal of Comparative Physiology A* 188: 663–681.

Iriarte, J. A., et al. 1990. Biogeographic variation of food habits and body size of American Puma. *Oecologia* 85: 185–190.

Klein, R. G. 1986. Carnivore size and Quaternary climate change in southern Africa. *Quaternary Research* 26: 153–170.

Londei, T. 2000. The cheetah (*Acinoyx jubatus*) dewclaw: Specialization overlooked. *Journal of Zoology, London* 251: 535–547.

Lorenz, K. 1988. *Man Meets Dog.* New York: Penguin Books.

McNab, B. K. 2002. *The Physiological Ecology of Vertebrates, a View from Energetics.* Ithaca: Cornell University Press.

Meiri, S., and T. Dayan. 2003. On the validity of Bergmann's rule. *Journal of Biogeography* 30: 331–351.

Miller, P. E. 2003. Vision in animals–what do dogs and cats see? *Waltham/Ohio State University Symposium,* www.vin.com/VINDBPub/SearchPB/Proceedings/PR05000/PR00510.htm.

Mugaas, J. N., and J. Seidensticker. 1993. Geographic variation of lean body mass and a model of its effect on the capacity of the raccoon to fatten and fast. *Bulletin of the Florida Museum of Natural History, Biological Series* 36: 85–107.

O'Brien, S. J., et al. 2002. The Feline Genome Project. *Annual Review of Genetics* 36: 657–686.

Ortolani, A., and T. M. Caro. 1996. The adaptive significance of color patterns in carnivores: Phylogenetic tests of classical hypotheses. Pp. 132–188 in J. L. Gittleman, ed., *Carnivore Behavior, Ecology, and Evolution,* vol. 2. Ithaca: Cornell University Press.

Quammen, D. 1998. The white tigers of Cincinnati. Pp. 81–89 in *Wild Thoughts from Wild Places.* New York: Simon & Schuster.

Radinsky, L. 1975. Evolution of the felid brain. *Brain, Behaviour and Evolution* 11: 214–245.

Ralls, K. 1978. When bigger is better. *New Scientist.* 9 February: 360–362.

Salles, J. F. O. 1992. Felid phylogenetics: Extant taxa and skull morphology (Felidae, Aeluroidea). *American Museum Novitates* 3047: 1–67.

Schmidt-Nielsen, K. 1984. *Scaling: Why Is Animal Size So Important?* Cambridge: Cambridge University Press.

Stanley, G. B., F. Li, and Y. Dan. 1999. Reconstruction of natural scenes from ensemble responses in the lateral geniculate nucleus. *Journal of Neuroscience* 19: 8036–8042.

Turner, A. 1997. *The Big Cats and Their Fossil Relatives.* New York: Columbia University Press.

Van Valkenburgh, B., and C. B. Ruff. 1987. Canine tooth strength and killing behaviour in large carnivores. *Journal of Zoology, London* 212: 379–397.

Wen, G. Y., J. A. Sturman, and J. W. Shek. 1985. A comparative study of the tapetum, retina and skull of the ferret, dog and cat. *Laboratory Animal Science* 35: 200–210.

Werdelin, L. 1983. Morphological patterns in the skulls of cats. *Biological Journal of the Linnean Society* 19: 375–391.

Werdelin, L., and L. Olsson. 1997. How the leopard got its spots: A phylogenetic view of the evolution of felid coat patterns. *Biological Journal of the Linnean Society* 62: 383–400.

Diet and Predation

Alexander, R. M. 1992. *Exploring Biomechanics: Animals in Motion.* New York: Scientific American Library.

Allen, M. E., O. T. Oftedal, and D. J. Baer. 1996. Feeding and nutrition of carnivores. Pp. 139–147 in D. G. Kleiman, M. E. Allen, K. V. Thompson, and S. Lumpkin, eds., *Wild Mammals in Captivity.* Chicago: University of Chicago Press.

Allen, M. E., et al. 1995. Do maintenance energy requirements of felids reflect their feeding strategies? *Proceeding of the Annual Conference of the AZA Nutritional Advisory Group* 1: 97–103.

Anderson, C. R., and F. G. Lindzey. 2003. Estimating cougar predation rates from GPS location clusters. *Journal of Wildlife Management* 67: 307–316.

Anton, M., and A. Turner. 1993. *The Big Cats and Their Fossil Relatives.* New York: Columbia University Press.

Avenant, N. L., and J. A. J. Nel. 2002. Among habitat variation in prey availability and use by caracal. *Mammalian Biology* 67: 18–33.

Beier, P., D. Choate, and R. H. Barrett. 1995. Movement patterns of mountain lions during different behaviors. *Journal of Mammalogy* 76: 1056–1070.

Berger, J., and J. D. Wehausen. 1991. Consequences of a mammalian predator-prey disequilibrium in the Great Basin Desert. *Conservation Biology* 5: 244–248.

Bloch, K. 1995. Why tabby takes to tuna. *Harvard Magazine* September/October: 50–53.

Brand, C. J., and L. B. Keith. 1979. Lynx demography during a snowshoe hare decline in Alberta. *Journal of Wildlife Management* 43: 827–849.

Carbone, C., and J. L. Gittleman. 2002. A common rule for the scaling of carnivore density. *Science* 295: 2273–2276.

Carbone, C., et al. 1999. Energetic constraints on the diet of terrestrial carnivores. *Nature* 402: 286–288.

Caro, T. M. 1994. *Cheetahs of the Serengeti Plains: Group Living in an Asocial Species*. Chicago: University of Chicago Press.

Colinvaux, P. 1978. *Why Big Fierce Animals Are Rare*. Princeton: Princeton University Press.

Curio, E. 1976. *The Ethology of Predation*. New York: Springer-Verlag.

Delibes, M., A. Rodrigus, and P. Ferreras. 2000. *Action Plan for the Conservation of the Iberian Lynx in Europe* (Lynx pardinus). Council of Europe Publishing, Nature and Environment, No. 111.

Dunstone, N., et al. 2002. Spatial organization, ranging behavior and habitat use of the kodkod (*Oncifelis guigna*) in southern Chile. *Journal of Zoology, London* 257: 1–11.

Eisenberg, J. F., and P. Leyhausen. 1972. The phylogenesis of predatory behavior in mammals. *Zeitschrift Tierpsychologie* 30: 59–93.

Elliott, J. P., I. McT. Cowan, and C. S. Holling. 1977. Prey capture by the African lion. *Canadian Journal of Zoology* 55: 1811–1828.

Emmons, L. H. 1988. A field study of ocelot (*Felis pardalis*) in Peru. *Review of Ecology* 43: 133–157.

Emmons, L. H., et al. 1989. Ocelot behavior in moonlight. *Advances in Neotropical Mammalogy* 1989: 233–242.

Fitzgerald, B. M. 1988. Diet of domestic cats and their impact on prey populations. Pp. 123–147 in D. C. Turner and P. Bateson, eds., *The Domestic Cat: The Biology of Its Behaviour*. Cambridge: University of Cambridge Press.

Geertsema, A. A. 1985. Aspects of the ecology of serval *Leptailurus serval* in the Ngorongoro Crater, Tanzania. *Netherlands Journal of Zoology* 35: 527–610.

Geist, V. 1998. *Deer of the World*. Mechanicsburg, PA: Stackpole Books.

Hanby, J. P., J. Bygott, and C. Packer. 1995. Ecology, demography, and behavior of lions in two contrasting habitats: Ngorongoro Crater and Serengeti Plains. Pp. 315–331 in A. R. E. Sinclair and P. Arcese, eds., *Serengeti II*. Chicago: University of Chicago Press

Henry, J. D. 1996. *Red Fox: The Catlike Canine*. Washington, DC: Smithsonian Institution Press.

Hornocker, M. G. 1970. An analysis of mountain lion predation upon mule deer and elk in the Idaho Primitive Area. *Wildlife Monographs* 21: 1–39.

Jedrzejewski, W., et al. 1993. Foraging by lynx and its role in ungulate mortality: The local (Bialowieza forest) and Palaearctic viewpoints. *Acta Theriologica* 38: 385–403.

Johnson, W. E., and W. L. Franklin. 1991. Feeding and spatial ecology of *Felis geoffroyi* in southern Patagonia. *Journal of Mammalogy* 72: 815–820.

Kantorosinski, S., and W. B. Morrison. 1988. A review of feline nutrition. *Iowa State University Veterinarian* 50: 95–106.

Karanth, U. K., and M. Sunquist. 2000. Behavioural correlates of predation by tiger (*Panthera tigris*), leopard (*Panthera pardas*), and dhole (*Cuon alpinus*). *Journal of Zoology, London* 250: 255–265.

Kitchener, A. 1991. *The Natural History of Wild Cats*. Ithaca, NY: Cornell University Press.

Knick, S. T. 1990. Ecology of bobcats relative to exploitation and prey decline in southeastern Idaho. *Wildlife Monographs* 108: 1–42.

Koehler, G. 1988. Bobcat bill of fare. *Natural History* 12(88): 48–56.

Krebs, C. J., et al. 1995. Impact of food and predation on the snowshoe hare cycle. *Science* 269: 1112–1115.

Kruuk, H. 1972. Surplus killing by carnivores. *Journal of Zoology, London* 166: 233–244.

———. 1986. Interactions between Felidae and their prey. Pp. 353–374 in S. D. Miller and D. D. Everett, eds., *Cats of the World: Biology, Conservation, and Management*. Washington, DC: National Wildlife Federation.

———. 1995. *Otters: Predation and Populations*. Oxford: Oxford University Press.

Leigh, E. J. 2002. *A Magic Web: The Forest of Barro Colorado Island*. Oxford: Oxford University Press.

Leyhasuen, P. 1979. *Cat Behavior*. New York: Garland STPM Press.

Ludlow, M. E., and M. E. Sunquist. 1987. Ecology and behavior of ocelots in Venezuela. *National Geographic Research* 3: 447–461.

MacDonald, L. M., Q. R. Rogers, and J. G. Morris. 1984. Nutrition of the domestic cat, a mammalian carnivore. *Annual Review of Nutrition* 4: 521–562.

Miquelle, D. C., et al. 1999. Hierarchical spatial analysis of Amur tiger relationships to habitat and prey. Pp. 71–99 in J. Seidensticker, S. Christie, and P. Jackson, eds., *Riding the Tiger: Tiger Conservation in Human-dominated Landscapes*. Cambridge: Cambridge University Press.

Morse, D. H. 1980. *Behavioral Mechanisms in Ecology*. Cambridge, MA: Harvard University Press.

Nellis, C. H., and L. B. Keith. 1968. Hunting activities and success of lynxes in Alberta. *Journal of Wildlife Management* 32: 718–722.

Okarma, H., et al. 1997. Predation of Eurasian lynx on roe deer and red deer in Bialowieza Primeval Forest. *Acta Theriologica* 42: 203–224.

Palomares, F., et al. 2001. Spatial ecology of Iberian lynx and abundance of European rabbits in southwestern Spain. *Wildlife Monographs* 148: 1–36.

Powell, R. A. 1993. *The Fisher: Life History, Ecology, and Behavior*. 2d ed. Minneapolis: University of Minnesota Press.

Price, B. M., V. C. Bleich, and R. T. Bowyer. 2000. Social organization of mountain lions: Does a land-tenure system regulate population size? *Ecology* 81: 1522–1543.

Robbins, C. T. 1983. *Wildlife Feeding and Nutrition*. New York: Academic Press.

Ruggiero, L. F., et al. 2002. *Ecology and Conservation of Lynx in the United States*. Fort Collins: University Press of Colorado and USDA Rocky Mountain Experiment Station.

Sanderson, J., M. E. Sunquist, and A. W. Iriate. 2002. Natural history and landscape-use of guigna (*Oncifelis guigna*) on the Isla Grande de Chiloe, Chile. *Journal of Mammalogy* 83: 608–613.

Scheel, D. 1993. Profitability, encounter rates, and prey choice of African lions. *Behavioral Ecology* 4: 90–97.

Schmidt, K. 1999. Variation in daily activity of the free-living Eurasian lynx (*Lynx lynx*) in Bialowieza Primeval Forest, Poland. *Journal of Zoology, London* 249: 417–425.

Seidensticker, J. 1976. On the ecological separation between tigers and leopards. *Biotropica* 8: 225–234.

———. 1983. Predation by *Panthera* cats and measures of human influence in habitats of south Asian monkeys. *International Journal of Primatology* 4: 323–326.

———. 2002. Tigers: Top carnivores in Asian tropical forests. Pp. 56–59 in E. Wikramanayake et al., eds. *Terrestrial Ecoregions of the Indo-Pacific: A Conservation Assessment*. Washington DC: Island Press.

Seidensticker, J., and C. McDougal. 1993. Tiger predatory behaviour, ecology and conservation. *Symposia of the Zoological Society of London* 65: 105–125.

Seidensticker, J., et al. 1973. Mountain lion social organization in the Idaho Primitive Area. *Wildlife Monographs* 35: 1–35.

Sliwa, A. 1994. Diet and feeding behavior of the black-footed cat (*Felis nigripes* Burchell, 1824) in Kimberley Region, South Africa. *Der Zoologische Garten* 64: 83–96.

Soule, M., and J. Terborgh. 1999. *Continental Conservation: Scientific Foundations of Regional Reserve Networks*. Washington DC: Island Press.

Stander, P. E. 1997. The ecology of asociality in Namibian leopards. *Journal of Zoology, London* 242: 343–364.

Sunquist, M. 1981. The social organization of tigers in Royal Chitawan National Park. *Smithsonian Contribution to Zoology* 336: 1–98.

Sunquist, M., and F. Sunquist. 1989. Ecological constraints on predation by large felids. Pp. 283–301 in J. L. Gittleman, ed., *Carnivore Behavior, Ecology, and Evolution*. Ithaca: Cornell University Press.

Terborgh, J., et al. 2001. Ecological meltdown in predator-free forest fragments. *Science* 294: 1923–1926.

Van Valkenburgh, B. 1989. Carnivore dental adaptations and diet: A summary of trophic diversity within guilds. Pp. 410–436 in J. L. Gittleman, ed., *Carnivore Behavior, Ecology, and Evolution*. Ithaca: Cornell University Press.

Social Behavior

Biben, M. 1979. Predation and predatory play behaviour of domestic cats. *Animal Behaviour* 27: 81–94.

Caro, T. M. 1989. Determinants of asociality in felids. Pp. 41–74 in V. Standen and R. A. Foley, eds., *Comparative Sociology: The Behavioural Ecology of Humans and Other Mammals*. Oxford: Blackwell Publishers.

———. 1994. *Cheetahs of the Serengeti Plains: Group Living in an Asocial Species*. Chicago: University of Chicago Press.

Caro, T., and M. D. Hauser. 1992. Is there teaching in nonhuman animals? *Quarterly Review of Biology* 67: 151–174.

Dunstone, N., et al. 2002. Social organization, ranging behaviour and habitat use of the kodkod (*Oncifelis guigna*) in southern Chile. *Journal of Zoology, London* 257: 1–11.

Fernandez, N., and F. Palomares. 2000. The selection of breeding dens by the endangered Iberian lynx (*Lynx pardinus*): Implications for its conservation. *Biological Conservation* 94: 51–61.

Gompper, M. E., and J. L. Gittleman. 1991. Home range scaling: Intraspecific and comparative trends. *Oecology* 87: 343–348.

Gorman, M. L., and B. J. Trowbridge. 1989. The role of odor in the social lives in the carnivores. Pp. 57–87 in J. L. Gittleman, ed., *Carnivore Behavior, Ecology, and Evolution*. Ithaca: Cornell University Press.

Grigione, M. M., et al. 2002. Ecological and allometric determinants of home-range size for mountain lions (*Puma concolor*). *Animal Conservation* 5: 317–324.

Grinnell, J. 1997. The lion's roar: More than just hot air. *ZooGoer* 26(3): 6–13.

Grinnell, J., and K. McComb. 2001. Roaring and social communication in African lions: The limitations imposed on listeners. *Animal Behaviour* 62: 93–98.

Harestad, A. A., and F. L. Bunnell. 1979. Home range and body weight—a reevaluation. *Ecology* 60: 389–402.

Hast, M. H. 1989. The larynx of roaring and non-roaring cats. *Journal of Anatomy* 163: 117–121.

Kays, R. W., and B. D. Patterson. 2002. Mane variation in African lions and its social correlates. *Canadian Journal of Zoology* 80: 471–478.

Kerley, L. L., et al. 2003. Reproductive parameters of wild female Amur (Siberian) tigers. *Journal of Mammalogy* 84: 288–298.

Kitchner, A. C. 1999. Watch with mother: A review of social learning in the Felidae. *Symposia of the Zoological Society of London* 72: 236–258.

Kleiman, D. G., and J. F. Eisenberg. 1973. Comparison of canid and felid social systems from an evolutionary perspective. *Animal Behaviour* 21: 637–659.

Martin, P., and P. Bateson. 1988. Behavioural development in the cat. Pp. 9–22 in D. C. Turner and P. Bateson, eds., *The Domestic Cat: The Biology of Its Behaviour*. Cambridge: University of Cambridge Press.

Mellen, J. D. 1993. A comparative analysis of scent marking, social and reproductive behavior in 20 species of small cats (*Felis*). *American Zoologist* 33: 151–166.

Morton, E. S., and J. Page. 1992. *Animal Talk*. New York: Random House.

Packer, C. 1986. The ecology of sociality in felids. Pp. 429–527 in D. I. Rubenstein and R. W. Wrangham, eds., *Ecological Aspects of Social Evolution*. Princeton: Princeton University Press.

Packer, C., and A. E. Pusey. 1997. Divided we call: Cooperation among lions. *Scientific American* 5: 52–59.

Peters, G. 1984. On the structure of friendly close range vocalizations in terrestrial carnivores (Mammalia: Carnivora: Fissipedia). *International Journal of Mammalian Biology* 49: 157–182.
———. 2002. Purring and similar vocalizations in mammals. *Mammal Review* 32: 245–271.

Peters, G., and M. H. Hast. 1994. Hyoid structure, laryngeal anatomy, and vocalizations in felids (Mammalia: Carnivora: Felidae). *International Journal of Mammalian Biology* 59: 87–104.

Ruiz-Miranda, C. R., et al. 1998. Vocalizations and other behavioral responses of male cheetahs (*Acinonyx jubatus*) during experimental separation and reunion trials. *Zoo Biology* 17: 1–16.

Seidensticker, J. 1977. Notes on the early maternal behavior of the leopard. *Mammalia* 41: 111–113.

Seidensticker, J., S. Christie, and P. Jackson, eds. 1999. *Riding the Tiger: Tiger Conservation in Human-dominated Landscapes*. Cambridge: Cambridge University Press.

Seidensticker, J., et al. 1973. Mountain lion social organization in the Idaho Primitive Area. *Wildlife Monographs* 35: 1–35.

Schmidt, K. 1998. Maternal behavior and juvenile dispersal in the Eurasian lynx. *Acta Theriologica* 43: 391–408.

von Schantz, T. 1984. Spacing strategies, kin selection, and population regulation in altricial vertebrates. *Oikos* 42: 48–58.

Walsh, E. J., et al. 2003. Acoustical communication in *Panthera tigris*: A study of tiger vocalization and auditory receptivity. *Acoustical Society of America ICA/ASA '03 Lay Language Papers*.

Wemmer, C., and K. Scow. 1977. Communication in the Felidae with emphasis on scent marking and contact patterns. Pp. 749–766 in T. A. Sebeok, ed., *How Animals Communicate*. Bloomington: Indiana University Press.

West, P. M., and C. Packer. 2002. Sexual selection, temperature, and the lion's mane. *Science* 297: 1339–1343.

Cat Life

Cherfas, J. 1987. How to thrill your cat this Christmas. *New Scientist* 24 December: 42–45.

Gittleman, J. L. 1986. Carnivore brain size, behavioral ecology, and phylogeny. *Journal of Mammalogy* 67: 23–36.

Kavanau, J. L. 1997. Origin and evolution of sleep: Roles of vision and endothermy. *Brain Research Bulletin* 42: 245–264.

Lawton, G. 2003. To sleep, perchance to dream. *New Scientist* 28 June: 28–35.

Lorenz, K. 1954. *Man Meets Dog*. New York: Penguin Books.

Maehr, D. S. 1997. *The Florida Panther: Life and Death of a Vanishing Carnivore*. Washington, DC: Island Press.

Murray, D. L., et al. 1999. Infectious disease and the conservation of free-ranging large carnivores. *Animal Conservation* 2: 241–254.

Packert, C. 1998. Why menopause? *Natural History* 107(6): 24–26.

Palomares, F., and T. M. Caro. 1999. Interspecific killing among mammalian carnivores. *American Naturalist* 153: 492–508.

2. Cat Evolution and Diversity
Evolution
Antone, M., and A. Turner. 1993. *The Big Cats and Their Fossil Relatives*. New York: Columbia University Press.

Anyonge, W. 1993. Body mass in large extant and extinct carnivores. *Journal of Zoology, London* 231: 339–350.

Eisenberg, J. F. 1981. *The Mammalian Radiations*. Chicago: University of Chicago Press.

Emerson, S. B., and L. Radinsky. 1980. Functional analysis of sabertooth cranial morphology. *Paleobiology* 6: 295–312.

Janczewski, D., et al. 1992. Molecular phylogenetic inference from saber-toothed cat fossils of Rancho La Brea. *Proceeding of the National Academy of Science* 89: 9769–9773.

Martin, L. D. 1980. Functional morphology and the evolution of cats. *Transactions of the Nebraska Academy of Science* 8: 141–154.

Van Valkenburgh, B. 1991. Cats in communities: Past and present. P. 16 in J. Seidensticker and S. Lumpkin, eds., *Great Cats*. Emmaus, PA: Rodale Press.

———. 1999. Major patterns in the history of carnivorous mammals. *Annual Review of Earth and Planetary Science* 27: 463–493.

Distribution and Abundance
Koehler, G., and M. G. Hornocker. 1991. Seasonal resource use among mountain lions, bobcats and coyotes. *Journal of Mammalogy* 72: 391–396.

Olson, D. M., et al. 2001. Terrestrial ecoregions of the world: A new map of life on Earth. *BioScience* 51: 933–938.

Ricklefs, R. E., and D. Schluter, eds. 1993. *Species Diversity in Ecological Communities: Historical and Geographical Perspectives*. Chicago: University of Chicago Press.

Seidensticker, J. 1976. On the ecological separation between tigers and leopards. *Biotropica* 8: 225–234.

Seidensticker, J., J. F. Eisenberg, and R. Simons. 1984. The Tangjiahe, Wanglang, and Feng-tongzhai giant panda reserve and biological conservation in the People's Republic of China. *Biological Conservation* 28: 217–251.

Soulé, M., ed. 1986. *Conservation Biology: The Science of Scarcity and Diversity*. Sunderland, MA: Sinauer Associates.

Diversity
Bininda-Emonds, O. R. P., J. L. Gittleman, and A. Purvis. 1999. Building large trees by compiling phylogenetic information: A complete phylogeny of the extant Carnivora (Mammalia). *Biological Review* 74: 141–175.

Johnson, W. E., E. Eizirik, et al. 1996. Resolution of recent radiations within three evolutionary lineages of Felidae using mitochrondrial restriction fragment length polymorphism variation. *Journal of Mammalian Evolution* 3: 97–120.

Johnson, W. E., P. A. Dratch, et al. 2001. Application of genetic concepts and molecular methods to carnivore conservation. Pp. 335–358 in J. L. Gittleman et al., eds., *Carnivore Conservation*. Cambridge: Cambridge University Press.

Mattern, M. Y., and D. A. McLennan. 2000. Phylogeny and speciation of felids. *Cladistics* 16: 232–253.

Mayr, E. 2001. *What Evolution Is*. New York: Basic Books.

Soulé, M., ed. 1986. *Conservation Biology: The Science of Scarcity and Diversity*. Sunderland, MA: Sinauer Associates.

Austin, S. A. 2002. *Ecology of Sympatric Carnivores in Khoa Yai National Park, Thailand*. Ph.D. dissertation, Texas A&M University, Kingsville.

Bailey, T. N. 1993. *The African Leopard: Ecology and Behavior of a Solitary Leopard*. New York: Columbia University Press.

Bertram, B., 1998. *Lions*. Stillwater, MN: Voyageur Press.

Daniel, J. C. 1996. *The Leopard in India*. Dehra Dun, India: Natraj Publishers.

Eizirik, E., et al. 2001. Phylogeography, population history and conservation genetics of jaguars (*Panthera onca*, Mammalia, Felidae). *Molecular Ecology* 10: 65–79.

Fox, J. L. 1989. *A Review of the Status and Ecology of the Snow Leopard* (Panthera uncia). Seattle: International Snow Leopard Trust.

Harvey, C., and P. Kat. 2000. *Prides: The Lions of Moremi*. Washington, DC: Smithsonian Institution Press.

Hemmer, H. 1972. *Uncia uncia. Mammalian Species* 20: 1–5.

Hoogesteijn, R., and E. Mondolfi. 1993. *The Jaguar*. Caracas: Armitano Publishers.

Jackson, R., and G. Ahlborn. 1989. Snow leopards in Nepal—home range and movement. *National Geographic Research* 5: 161–175.

Jenny, D. 1996. Spatial organization of leopard *Panthera pardus* in Tai National Park, Ivory Coast: Is rainforest habitat a "tropical haven"? *Journal of Zoology, London* 240: 427–440.

Karanth, K. U. 2001. *The Way of the Tiger*. Stillwater, MN: Voyageur Press.

Kingdon, J. 1977. *East African Mammals*, vol. 3A. Chicago: University of Chicago Press.

Kurten, B., and E. Anderson. 1980. *Pleistocene Mammals of North America*. New York: Columbia University Press.

Lekagul, B., and J. A. NcNeely. 1977. *Mammals of Thailand*. Bangkok: Association for the Conservation of Wildlife.

Logan, K. A., and L. L. Sweanor. 2001. *Desert Puma: Evolutionary Ecology and Conservation of an Enduring Carnivore*. Washington, DC: Island Press.

Mazak, V. 1981. *Panthera tigris. Mammalian Species* 152: 1–8.

Miththapala, S., J. Seidensticker, and S. J. O'Brien. 1996. Phylogeographic subspecies recognition in leopard (*Panthera pardus*): Molecular genetic variation. *Conservation Biology* 10: 1115–1132.

Nunez, R., B. Miller, and F. Lindzey. 2000. Food habits of jaguars and pumas in Jalisco, Mexico. *Journal of Zoology, London* 252: 373–379.

Packer, C., and A. E. Pusey. 1997. Divided we fall: Cooperation among lions. *Scientific American* 1997(5): 52–59.

Rabinowitz, A. 2000. *Jaguar*. Washington, DC: Island Press.

Schaller, G. B. 1967. *The Deer and the Tiger*. Chicago: University of Chicago Press.

———. 1972. *The Serengeti Lion.* Chicago: University of Chicago Press.

Schaller, G. B., and P. G. Crawshaw. 1980. Movement patterns of jaguar. *Biotropica* 12: 161–168.

Seidensticker, J. 1996. *Tigers*. Stillwater, MN: Voyageur Press.

Seidensticker, J., S. Christie, and P. Jackson, eds. 1999. *Riding the Tiger: Tiger Conservation in Human-dominated Landscapes*. Cambridge: Cambridge University Press.

Seymour, K. L. 1989. *Panthera onca. Mammalian Species* 340: 1–9.

Stander, P. E., et al. 1997. The ecology of sociality in Namibian leopards. *Journal of Zoology, London* 242: 343–364.

Sunquist, M. E. 1981. The social organization of tigers in Royal Chitawan National Park. *Smithsonian Contribution to Zoology* 336: 1–98.

Thapar, V. 1988. *Tigers, the Secret Life*. Emmaus, PA: Rodale Press.

Uphyrkina, O., et al. 2001. Phylogenetics, genome diversity and origin of modern leopard, *Panthera pardus*. *Molecular Ecology* 10: 2617–2633.

LYNX LINEAGE

Bailey, T. N. 1974. Social organization in a bobcat population. *Journal of Wildlife Management* 38: 435–446.

Breitenmoser, P., et al. 1993. Spatial organization and recruitment of lynx (*Lynx lynx*) in a reintroduced population in the Swiss Jura Mountains. *Journal of Zoology, London* 321: 449–464.

Delibes, M., A. Rodrigues, and P. Ferreras. 2000. *Action Plan for the Conservation of the Iberian Lynx in Europe* (Lynx pardinus). Council of Europe Publishing, Nature and Environment, No. 111.

Ferreas, P., et al. 1997. Spatial organization and land tenure of the endangered Iberian lynx (*Lynx pardinus*). *Journal of Zoology, London* 243: 63–189.

Halanycch, K. M., et al. 1999. Cytochrome *b* phylogeny of North American hares and jackrabbits (*Lepus*, Lagomorpha) and the effects of saturation in outgroup taxa. *Molecular Phylogenetics and Evolution* 11: 213–221.

Larviere, S., and L. Walton. 1997. *Lynx rufus*. *Mammalian Species* 563: 1–8.

Nellis, C., and L. B. Keith. 1968. Hunting activities and success of lynxes in Alberta. *Journal of Wildlife Management* 32: 718–722.

Ruggiero, L. F., et al. 2000. *Ecology and Conservation of Lynx in the United States*. Boulder: University of Colorado Press.

Tumlison, R. 1987. *Felis lynx*. *Mammalian Species* 269: 1–8.

Young, S. P. 1958. *The Bobcat of North America*. Lincoln: University of Nebraska Press.

LEOPARD CAT LINEAGE AND RUSTY-SPOTTED CAT

Carlstead, K., J. L. Brown, and J. Seidensticker. 1993. Behavioral and adrenocortical responses to environmental changes in leopard cats (*Felis bengalensis*). *Zoo Biology* 12: 321–331.

DeAlwis, W. F. 1973. Status of Southeast Asia's small cats. Pp. 198–208 in R. L. Eaton, ed., *The World's Cats*. Vol. 1, *Ecology and Conservation*. Winston, OR: World Wildlife Safari.

Grassman, L. I. 1998. Movements and prey selection of leopard cats (*Prionailurus bengalensis*). *Societa Zoological la Torbiera Scientific Report* 4: 5–12.

Heptner, V. G., and A. A. Sludskii, eds. 1992. *Mammals of the Soviet Union*. Vol. 2, part 2, *Carnivora (Hyaenas and Cats)*. Washington, DC: Smithsonian Institution Libraries and the National Science Foundation.

Muul, I., and B. L. Lim. 1970. Ecological and morphological observation of *Felis planiceps*. *Journal of Mammalogy* 51: 806–808.

Rabinowitz, A. 1990. Notes on the behavior and movement of leopard cats, *Felis bengalensis*, in a dry tropical forest mosaic in Thailand. *Biotropica* 22: 397–403.

Rajaratnam, R. 2000. Ecology of the Leopard Cat (*Prionailurus bengalensis*) in Tabin Wildlife Reserve. Ph.D. dissertation, University of Kebangsaan, Bangi, Malaysia.

Seidensticker J. 2003. Fishing cats enjoy city life. http://nationalzoo.si.edu/ConservationAndScience/SpotlightOnScience/seidenstickerj20030526.cfm.

Shepherdson, D., et al. 1993. The influence of food presentation on the behavior of small cats in confined environments. *Zoo Biology* 12: 203–216.

Yasuma, S. 1988. Iriomote cat: King of the night. *Animal Kingdom* 91(6): 12–21.

CARACAL LINEAGE AND SERVAL

Geertsema, A. A. 1985. Aspects of the ecology of the serval *Leptailurus serval* in the Ngorongoro Crater, Tanzania. *Netherlands Journal of Zoology* 35: 527–610.

Kingdon, J. 1977. *East African Mammals*, vol. 3A. Chicago: University of Chicago Press.

Van Mensch, P. J. A., and P. J. H. Van Bree. 1969. On the African golden cat, *Profelis aurata* (Temminck, 1827). *Biologia Gabonica* 5: 235–269.

BAY CAT LINEAGE

Dabrowska, A., and J. Smielowski. 2001. Some observations on the behavior of the Chinese golden cat, *Catopuma temmincki tristis* (Milne-Edwards, 1872) at Wassenaar Wildlife Breeding Center. *Der Zoologische Garten* 71: 394–402.

Lekagule, B., and J. A. NcNeely. 1977. *Mammals of Thailand*. Bangkok: Association for the Conservation of Wildlife.

Sunquist, M., et al. 1994. Rediscovery of the Bornean bay cat. *Oryx* 28: 67–70.

PUMA LINEAGE

Anderson, A. E. 1983. A critical review of literature on puma (*Felis concolor*). *Colorado Division of Wildlife Special Report* 54: 1–91.

Beier, P., D. Choate, and R. H. Barrett. 1995. Movement patterns of mountain lions during different behaviors. *Journal of Mammalogy* 76: 1056–1075.

Caro, T. M. 1994. *Cheetahs of the Serengeti Plains: Group Living in an Asocial Species*. Chicago: University of Chicago Press.

Culver, M., et al. 2000. Genomic ancestry of the American puma. *Journal of Heredity* 91: 186–197.

Currier, M. J. P. 1983. *Felis concolor*. *Mammalian Species* 200: 1–7.

de Oliveira, T. G. 1998. *Herpailurus yagouaroundi*. *Mammalian Species* 578: 1–6.

Konecny, M. J. 1989. Movement patterns and food habits of four sympatric carnivore species in Belize, Central America. *Advances in Neotropical Mammalogy* 1989: 243–264.

Laurenson, K. M. 1994. High juvenile mortality in cheetahs (*Acinonyx jubatus*) and its consequence for maternal care. *Journal of Zoology, London* 234: 387–408.

Logan, K. A., and L. L. Sweanor. 2001. *Desert Puma*. Washington, DC: Island Press.

Maehr, D. S. 1997. *The Florida Panther: Life and Death of a Vanishing Carnivore*. Washington, DC: Island Press.

Seidensticker, J., et al. 1973. Mountain lion social organization in the Idaho Primitive Area. *Wildlife Monographs* 35: 1–35.

Young, S. P., and E. A. Goldman. 1964. *The Puma: Mysterious American Cat*. New York: Dover Publications.

FELIS LINEAGE AND PALLAS' CAT

Bennett, S. W., and J. D. Mellen. 1983. Social interactions and solitary behaviors in a pair of captive sand cats (*Felis margarita*). *Zoo Biology* 2: 39–46.

Daniels, M., et al. 1998. Morphological and pelage characteristics of wild living cats in Scotland: Implication for defining the "wildcat." *Journal of Zoology, London* 244: 231–247.

Dragesco-Joffe, A. 1993. *La vie sauvage au Sahara*. Paris: Delachaux et Niestle.

Groves, C. P. 1980. The Chinese mountain cat. *Carnivore* 3(3): 35–41.

Heptner, V. G., and A. A. Sludskii, eds. 1992. *Mammals of the Soviet Union*. Vol. 2, part 2, *Carnivora (Hyaenas and Cats)*. Washington, DC: Smithsonian Institution Libraries and the National Science Foundation.

Kingdon, J. 1977. *East African Mammals*, vol. 3A. Chicago: University of Chicago Press.

Roberts, T. J. 1977. *The Mammals of Pakistan*. London: Ernest Benn.

Sausman, K. A. 1997. Notes on the history of sand cats, *Felis margarita*, in captivity. *Der Zoologische Garten* 67: 81–84.

Sliwa, A. 1994. Diet and feeding behavior of the black-footed cat (*Felis nigripes* Burchell, 1924) in the Kimberley Region, South Africa. *Der Zoologische Garten* 64: 83–96.

Wiseman, R., C. O'Ryan, and E. H. Harley. 2000. Microsatellite analysis reveals that domestic cat (*Felis catus*) and southern African wild cat (*F. lybica*) are genetically distinct. *Animal Conservation* 3: 221–228.

De la Rosa, C. L., and C. C. Nocke. 2000. *A Guide to the Carnivores of Central America*. Austin: University of Texas Press.

de Oliveira, T. G. 1998. *Leopardus wiedii. Mammalian Species* 579: 1–6.

Eisenberg, J. F., and K. R. Redford. 1999. *Mammals of the Neotropics*. Vol. 3, *The Central Neotropics: Ecuador, Peru, Bolivia, and Brazil*. Chicago: University of Chicago Press.

Eizirik, E. 1998. Phylogeographic patterns and evolution of the mitochondrial DNA control region in two neotropical cats (Mammalia, Felidae). *Journal of Molecular Evolution* 47: 613–624.

Emmons, L. 1988. A field study of ocelots (*Felis pardalis*) in Peru. *Review of Ecology* 43: 133–157.

Garcia-Perea, R. 1994. Pampas cats: How many species? *Cat News* 20: 21–24.

Johnson, W. E., and W. L. Franklin. 1991. Feeding and spatial ecology of *Felis geoffroyi* in southern Patagonia. *Journal of Mammalogy* 72: 815–820.

Johnson, W. E., M. Culver, et al. 1998. Tracking the evolution of the elusive Andean Mountain cat (*Oreailurus jacobita*) from mitochondrial DNA. *Journal of Heredity* 89: 227–232.

Johnson, W. E., J. Pecon-Slattery, et al. 1999. Disparate phylogeographic patterns of molecular genetic variation in four closely related South American small cat species. *Molecular Ecology* 8: 579–594.

Ludlow, M. E., and M. E. Sunquist. 1987. Ecology and behavior of ocelots in Venezuela. *National Geographic Research* 3: 447–461.

Murray, J. L., and A. L. Gardner. 1997. *Leopardus pardalis. Mammalian Species* 548: 1–10.

Peterson, M. K. 1978. Behavior of the margay. *Carnivore* 1: 87–92.

Redford, K. H., and J. F. Eisenberg. 1992. *Mammals of the Neotropics*. Vol. 2, *The Southern Cone: Chile, Argentina, Uruguay, Paraguay*. Chicago: University of Chicago Press.

Walker, S., and A. Novaros. 2003. *Second Report on the Multinational Initiative to Determine the Status of the Andean Mountain Cat and Priorities for Its Conservation*. Neuquen, Argentina: Wildlife Conservation Society.

Ximenex, A. 1975. *Felis geoffroyi. Mammalian Species* 54: 1–4.

Yensen, E., and K. L. Seymour. 2000. *Oreailurus jacobita. Mammalian Species* 644: 1–6.

3. Cats and Humans

Decline and Recovery

Ferguson, S. H., and S. Lariviere. 2002. Can comparing life histories help conserve carnivores? *Animal Conservation* 5: 1–12.

Gittleman, J .L., S. M. Funk, D. Macdonald, and R. K. Wayne, eds. 2001. *Carnivore Conservation*. Cambridge: Cambridge University Press.

Humphrey, S. R., and B. M. Stith. 1990. A balanced approach to conservation. *Conservation Biology* 4: 341–343.

IUCN. 2003. 2003 IUCN Red List of Threatened Species. www.redlist.org, downloaded on November 20, 2003.

Kellert, S. R. 1996. *The Value of Life: Biological Diversity and Human Society*. Washington, DC: Island Press.

Lumpkin, S., and J. Seidensticker. 2002. *Smithsonian Book of Giant Pandas*. Washington, DC: Smithsonian Institution Press.

Mace, G. M., A. Balmford, and J. R. Ginsberg, eds. 1998. *Conservation in a Changing World*. Cambridge: Cambridge University Press.

McKinney, M. L. 2001. Role of human population size in raising bird and mammal threats among nations. *Animal Conservation* 4: 45–57.

Nowell, K. 2000. *Far from a Cure: The Tiger Trade Revisited*. Cambridge: TRAFFIC International.

Sanderson, E. W., et al. 2002. The human footprint and the last of the wild. *BioScience* 52: 891–904.

Seidensticker, J., S. Christie, and P. Jackson, eds. 1999. *Riding the Tiger: Tiger Conservation in Human-dominated Landscapes.* Cambridge: Cambridge University Press.

Simberloff, D. 1998. Flagships, umbrellas, and keystones: Is single-species management passé in the landscape era? *Biological Conservation* 83: 247–257.

Wilcove, D. S., et al. 1998. Quantifying threats to imperiled species in the United States. *BioScience* 48: 607–615.

Woodroffe, R., and J. R. Ginsberg. 1998. Edge effects and the extinction of populations inside protected areas. *Science* 280: 2126–2128.

Domestic Cats

Budiansky, S. 2002. *The Character of Cats.* New York: Viking.

Clutton-Brock, J. 1988. *Cats: Ancient and Modern.* Cambridge, MA: Harvard University Press.

Diamond, J. 1997. *Guns, Germs, and Steel.* New York: Norton.

Keitt, B. S., et al. 2002. The effects of feral cats on the population viability of black-vented shearwaters (*Puffinus opisthomales*) on Natividad Island, Mexico. *Animal Conservation* 5: 217–223.

Lorenz, K. 1988. *Man Meets Dog.* New York: Penguin Books.

Rogers, K. M. 1998. *The Cat and the Human Imagination: Feline Images from Bast to Garfield.* Ann Arbor: University of Michigan Press.

Serpal, J. A. 1991. Domestic cats. Pp. 184–189 in J. Seidensticker and S. Lumpkin, eds., *Great Cats.* Emmaus, PA: Rodale Press.

Turner, D. C., and P. Bateson, eds. 2000. *The Domestic Cat: The Biology of Its Behaviour.* 2d ed. Cambridge: Cambridge University Press.

Man Killing

Beier, P. 1991. Cougar attacks on humans in the United States and Canada. *Wildlife Society Bulletin* 19: 403–412.

Berwick, S. 1978. The Gir Forest: An endangered ecosystem. *American Scientist* 64: 28–40.

Boomgaard, P. 2001. *Frontiers of Fear: Tigers and People in the Malay World, 1600–1950.* New Haven: Yale University Press.

Corbett, J. 1946. *Man Eaters of Kumaon.* Oxford: Oxford University Press.

Fitzhugh, E. L., and David P. Fjelline. 1997. Suggested responses to different puma behaviors. Pp. 26–28 in W. D. Padley, ed., *Proceedings of 5th Mountain Lion Workshop.* http://www.frii.com/~mytymyk/lions/pumadfn.htm.

Hoogesteijn, R., and E. Mondolfi. 1993. *The Jaguar.* Caracas: Armitano Publishers.

Kruuk, H. 2002. *Hunter and Hunted: Relationships between Carnivores and People.* Cambridge: Cambridge University Press.

O'Connell, J. F., K. Hawkes, and N. Burton Jones. 1988. Hadza scavenging: Implications for Plio/Pleistocene hominid subsistence. *Current Anthropology* 29: 356–363.

Patterson, B. D., E. J. Neiburger, and S. M. Kasiki. 2003. Tooth breakage and dental disease as a cause of carnivore-human conflicts. *Journal of Mammalogy* 84: 190–196.

Patterson, J. H. 1907. *The Man-eaters of Tsavo.* London: Macmillan.

Saberwal, V. K., et al. 1994. Lion-human conflict in the Gir Forest, India. *Conservation Biology* 8: 501–507.

Seidensticker, J. 2002. Tiger tracks. *Smithsonian* 32(10): 62–69.

Seidensticker, J., and S. Lumpkin. 1992. Mountains lions don't stalk people: True or false? *Smithsonian* 22(2): 113–122.

Seidensticker, J., et al. 1976. Problem tiger in the Sundarbans. *Oryx* 13: 267–273.

Seymour, K. L. 1989. *Panthera onca. Mammalian Species* 340: 1–9.

Treves, A., and L. Naughton-Treves. 1999. Risk and opportunity for human coexisting with large carnivores. *Journal of Human Evolution* 36: 275–282.

Cats and Culture

Darwin, C. 1845. *The Voyage of the Beagle*. Chap. 6, "Bahia Blanca to Buenos Aires," in *An Online Library of Literature*, http://www.literature.org/authors/darwin-charles/the-voyage-of-the-beagle/chapter-06.html.

Davidson, A. 1999. *The Oxford Companion to Food*. Oxford: Oxford University Press.

Lumpkin, S. 1991. Cats and culture. Pp. 190–203 in J. Seidensticker and S. Lumpkin, eds., *Great Cats*. Emmaus, PA: Rodale Press.

McNeely, J. A., and P. S. Wachtel. 1988. *Soul of the Tiger*. New York: Doubleday.

Matthiessen, P. 2000. *Tigers in the Snow*. New York: North Point Press.

O'Brien, S. J., and M. Dean. 1997. In search of AIDS-resistant genes. *Scientific American* 227(3): 44–51.

Rogers, K. M. 1998. *The Cat and the Human Imagination*. Ann Arbor: University of Michigan Press.

Cats and Science

Boitani, L., and T. K. Fuller. 2000. *Research Techniques in Animal Ecology: Controversies and Consequences*. New York: Columbia University Press.

Carbone, C., et al. 2001. The use of photographic rates to estimate densities of tigers and other cryptic mammals. *Animal Conservation* 4: 75–79.

Caro, T. M., and S. M. Durant. 1991. Use of qualitative analysis of pelage characteristics to reveal family resemblances in genetically monomorphic cheetahs. *Journal of Heredity* 82: 8–14.

Eldredge, N. 2000. *The Pattern of Evolution*. New York: W. H. Freeman.

Frank, L., D. Simpson, and R. Woodroffe. 2003. Foot snares: an effective method for capturing African lions. *Wildlife Society Bulletin* 32: 309–314.

Geertsema, A. A. 1985. Aspects of the ecology of serval *Leptailurus serval* in the Ngorongoro Crater, Tanzania. *Netherlands Journal of Zoology* 35: 527–610.

Goodrich, J. M., et al. 2001. Capture and chemical anesthesia of Amur (Siberian) tigers. *Wildlife Society Bulletin* 29: 533–542.

Kelly, M. J. 2001. Computer-aided photograph matching in studies using individual identification: An example from Serengeti cheetahs. *Journal of Mammalogy* 82: 440–449.

Leyhausen, P. 1965. The communal organization of solitary mammals. *Symposium of the Zoological Society of London* 14: 249–263.

Logan, K. A., L. L. Sweanor, J. F. Smith, and M. G. Hornocker. 1999. Capturing pumas with foot-hold snares. *Wildlife Society Bulletin* 27: 201–208.

Maehr, D. S. 1997. *The Florida Panther: Life and Death of a Vanishing Carnivore*. Washington, DC: Island Press.

Mayr, E. 1997. *This Is Biology: The Science of the Living World*. Cambridge, MA: Harvard University Press.

Meadows, R. 2002. Scat-sniffing dogs. *ZooGoer* 31(5): 22–27.

Mills, M. G. L. 1996. Methodological advances in capture, census, and food habits studies of large African Carnivores. Pp. 223–242 in J. L. Gittleman, ed., *Carnivore Behavior, Ecology, and Evolution*, vol. 2. Ithaca: Cornell University Press.

Miththapala, S., et al. 1989. Identification of individual leopards (*Panthera pardus kotiya*) using spot pattern variation. *Journal of Zoology, London* 218: 527–536.

O'Brien, S. J. 2003. *Tears of the Cheetah: And Other Tales from the Genetic Frontier*. New York: St. Martin's Press.

Packer, C. 1992. Captive in the wild. *National Geographic* 181(4): 122–136.

————. 1994. *Into Africa*. Chicago: University of Chicago Press.

Pennycuick, C., and J. Rudani. 1970. A method of identifying individual lions (*Panthera leo*) with an analysis of reliability of the identification. *Journal of Zoology, London* 160: 497–500.

Schaller, G. B. 1967. *The Deer and the Tiger*. Chicago: University of Chicago Press.

Seidensticker, J. 1996. *Tigers*. Stillwater, MN: Voyageur Press.

Seidensticker, J., S. Christie, and P. Jackson, eds. 1999. *Riding the Tiger: Tiger Conservation in Human-dominated Landscapes*. Cambridge: Cambridge University Press.

Seidensticker, J., et al. 1973. Mountain lion social organization in the Idaho Primitive Area. *Wildlife Monographs* 36: 1–63.

Smallwood, K. S., and E. L. Fitzhugh. 1993. A rigorous technique for identifying individual mountain lions *Felis concolor* by their tracks. *Biological Conservation* 65: 51–59.

Sunquist, F., and M. Sunquist. 1988. *Tiger Moon*. Chicago: University of Chicago Press.

Torelle, M., and M. Kery. 2003. Estimation of ocelot density in the Pantanal using capture-recapture analysis of camera-trapping data. *Journal of Mammalogy* 84: 607–614.

Zeldin, T. 2000. *Conversation*. New York: HiddenSpring.

TAXONOMIC INDEX

Taxa are indexed by their core common names; for example, "Asian golden cat" is listed as "golden cat, Asian." The scientific equivalent follows in parentheses. Page-number citations in *italics* refer to photographs or figures on those pages.

bay cat (*Catopuma badia*), 51, *134*, 135, 141, 143, *143*, 149–150, 155–156, 163, 169

black-footed cat (*Felis nigripes*), 19, 34, 37, 50–51, 59, 68, 75–76, 90, 92, 99, 104–105, 108, 111, 133, *134*, 141, 151–152

bobcat (*Lynx rufus*), 25, 38, 40, 45, 60, 67, 79, 90, 92, 99, 105, 107–108, 111–112, 121–123, 126, 128, *134*, 147–148, 156–157, 204, 216–218

caracal (*Caracal caracal*), 6, *18*, 25, 29, 38, 40, *44*, 45, 52, 68, 99, 103, 108, 121, *134*, 135, 141–142, 149, 169, 176, 202

cheetah (*Acinonyx jubatus*), 8, *11*, 12, *23*, 24, 25, 36, 38–39, 42–43, 45, 54, 57, 63, 66, 69, 69, *73*, 76, 78, 80, 84, 87, *92*, 95–97, 99, 101, 103–106, *106*, 108, 112–115, *113*, *115*, 121, 123, 125–126, 129, 133, *134*, 135, 142, 163, 168, 170, 176, 190, 192, 202, *202*, 204, 206, 214–216

clouded leopard (*Neofelis nebulosa*), 6, 10–12, 22, 25, 37, 51, 57–58, 95–96, 133, *134*, 137–138, 141, 144–146, 154–155, 157, 190–191, 202, 218

dawn cat (*Proailurus*), 132

domestic cat (*Felis catus*), 19, 25, 27–29, 34, 36–37, 52, 63, 66, 79, 81, 95, 98–99, 102–103, 105, 110–114, 117–118, 120–124, 127–129, *134*, 151, 160, 168, 176–189, *181*, *186–187*, 197, 203–206, 208–209, 221

fishing cat (*Prionailurus viverrinus*), 11, 46, 51, *53*, 54, 58, 84, 105, *134*, 135, 142, 148–149, 154, 156–157, 165, 168, 172, 216

flat-headed cat (*Prionailurus planiceps*), 6, 11, 34, 46, 51, *134*, 141, 148–149, 154, 156–157, 172

Geoffroy's cat (*Leopardus geoffroyi*), 11, 25, *26*, 59, 89, 92, *134*, 152–154, 184

golden cat
   African (*Profelis aurata*), 6, 51, *134*, 135, 141, 149, 163, 202, 204
   Asian (*Catopuma temminckii*), 6, 24, 51, 91, 99, *134*, 135, 141, 143, 149–150, 154–157

Iriomote cat (*Prionailurus iriomotensis*), 3, 143, 148–149

jaguar (*Panthera onca*), 22, 25, 27, 37–41, 47, 52, 57, *64*, *65*, 95–96, 99, 103, 121, 127, 129, *134*, 144–147, 150–151, 163, 170, 190, 200–201, 204, 217

jaguarundi (*Puma yagouaroundi*), 24, 26–27, 37, 51–52, 99, 105, *134*, 150, 163

jungle cat (*Felis chaus*), 45, 51, 59, 66, 121–122, *134*, 151–152, 154–155, 157, 165, 184

kodkod (*Leopardus guigna*), 25, 34, 59, 87, *134*, 141, 152–153

leopard (*Panthera pardus*), 12, 22, 24, 25–27, *26*, 31–32, 38–39, 42–43, *44*, 45, 50, 52, 54, 57, 60, 63, 67, 69, *73*, 74, 76, *77*, 78–80, 83, 91, 95–96, 99–100, 105, 108, 112, 121–124, 126–127, 129, 133, *134*, 139, 141, 144–146, 154–157, *162*, 163–164, 168, 172, 190–191, 196, 201–202, 204, 206, 213, 216–218
   African (*Panthera pardus pardus*), 22
   Amur (*Panthera pardus orientalis*), 22, 124
leopard cat (*Prionailurus bengalensis*), 11, 34, 38, 51, 54, 58, 60, 99, 112, *134*, 135, 141–142, 148–149, 154–157, 163, 165, 218
lion (*Panthera leo*), 8, 10, 12, 18–19, *20*, 22, *23*, 24, 25, 29–36, *30*, *33*, *35*, 38–39, 42, *44*, 45,

SUBJECT INDEX